The Biology
of Complex Organisms

Creation and Protection of Integrity

Edited by Klaus Eichmann

Birkhäuser Verlag
Basel · Boston · Berlin

Editor:

Prof. Dr. Klaus Eichmann
Max-Planck-Institut für Immunbiologie
Stübeweg 51
D–79108 Freiburg

Library of Congress Cataloging-in-Publication Data
The biology of complex organisms : creation and protection of integrity /
edited by Klaus Eichmann.
   p. cm.
   ISBN 3764369795 (alk. paper)
   1. Immune system--Congresses. 2. Evolution--Congresses. I. Eichmann, Klaus, 1939-
   QR180.3 .B53 2003
   571.9'6--dc21
                                     2002028140

Bibliographic information published by Die Deutsche Bibliothek
Die Deutsche Bibliothek lists this publication in the Deutsche Nationalbibliographie; detailed
bibliographic data is available in the internet at <http://dnb.ddb.de>.

ISBN 3-7643-6979-5 Birkhäuser Verlag, Basel - Boston - Berlin

© 2003 Birkhäuser Verlag, P.O. Box 133, CH-4010 Basel, Switzerland
Member of the BertelsmannSpringer Publishing Group
Printed on acid-free paper produced from chlorine-free pulp. TCF ∞
Cover illustration: Mouse embryo and antibody, logo of the Max-Planck-Institut für Immunbiologie
Cover design: Micha Lotrovsky, Therwil, Schweiz
Printed in Germany

ISBN 3-7643-6979-5

9 8 7 6 5 4 3 2 1

Table of Contents

List of authors

Prof. Dr. Dr. h.c. Klaus Eichmann
Director, Max-Planck-Institut für Immunbiologie
Stuebeweg 51, D-79108 Freiburg, Germany

Prof. Dr. Rudolf Jaenisch
Whitehead Institute for Biomedical Research
9 Cambridge Center, Cambridge, MA 02142, USA

Prof. Dr. Charles A. Janeway, Jr.
Sec Immunbiolology HHMI
Yale University School of Medicine
310 Cedar St, LH 416, New Haven, CT 06510, USA

Prof. Dr. Philippe Kourilsky
Institut Pasteur
Immunology Department
25, rue du Docteur Roux, F-75724 Paris cedex 15, France

Prof. Dr. J.F.A.P. Miller
The Walter and Eliza Hall Institute
PO Royal Melbourne Hospital, Melbourne, Victoria 3050, Australia

Prof. Dr. Hartmut H. Peter
Klinikum der Universität Freiburg
Abt. Rheumatologie u. klin.Immunologie
Hugstetter Str. 55, D-79106 Freiburg, Germany

Prof. Dr. Martin Raff
MRC Lab. for Molecular Cell
University College London
Gower Street, London WC1E 6BT, UK

Prof. Dr. Dr. h.c. Otto Westphal
Chemin de Ballalaz 18, CH-1820 Montreux, Switzerland

Prof. Dr. Lewis Wolpert
Royal Free and University College Medical School
University College London
Dept. of Anatomy and Devel. Biology
Gower Street, London WC1E 6BT, UK

Acknowledgement

This volume would not have been possible without the help of others within and without the Max-Planck-Institute. Particularly I would like to thank the speakers for editing the transcripts of their lectures and preparing them for written publication. I am indebted to Nele Leibrock for tireless work in preparing the text, to Lore Lay for preparing the figures, and to Christoph Gartmann and Stefan Kuppig for compiling the video presented on the DVD disk.

Klaus Eichmann
Freiburg, August 8, 2002

Preface

On December 6, 1961, a contract was signed by which the research institute of the Wander AG in Freiburg became the Max-Planck-Institut für Immunbiologie. The transfer of ownership took place during a happy expansion phase of the Max-Planck-Society in which a growing economy in Germany allowed the foundation of many new research institutes by the Max-Planck-Society and other organizations. Nevertheless, it was a remarkable event. The acquisition by an academic organization of an institute formerly operated by an industrial company was rather unusual, not to speak of the fact that not only the facilities but also the entire scientific personnel were taken over. Retrospectively, the 40 years of the institute in the Max-Planck-Society can be divided into 2 very different phases of 20 years each. The first 20 years were characterized by a continuation of the research that had begun in the Wander institute and centered on the structure and function of the bacterial compound endotoxin. During the second 20 years, the institute more than doubled in size and developed into an interdisciplinary research center that focuses on the development and organization of multicellular systems by combining studies in two fields of research: immunology and developmental biology.

The 40th anniversary of the foundation of the Max-Planck-Institute was celebrated by a ceremony including a scientific symposium. The first part of this volume presents the lectures given at the symposium by six leading biologists. They present their views on principles nature uses in the evolution and generation of complex organisms such as mice and man, and how the immune system manages to protect their integrity in a hostile environment. The lectures were initially not meant for written publication. All of the lectures, however, were very well received by the attendees not only because they were given in a style appreciated by a general audience. Although by no means covering the subjects in any comprehensive or complete sort of way, they nevertheless appear to present well focused and thoughtful accounts of some of the issues that are in the center of present day scientific and public interest. In addition, the lectures lend strong support to the philosophy of the institute that understanding complex organisms is one of the continuing challenges of biology and that the general principles that govern such organisms are best unraveled by studying multicellular systems in their various forms. As a result of the very positive feedback, it seemed worthwhile to compile the materials in a book.

This volume is a *Festschrift* celebrating the 40th anniversary of the Max-Planck-Institut für Immunbiology. Accordingly, the scientific lectures are complemented by historical accounts of the early and late phases of the Max-Planck-Institute and its standing within the history of immunological research in the city of Freiburg. In addition, the volume contains a DVD video diskette featuring the speakers in short sections of their talks, for those who are interested in the human individual behind the science.

Klaus Eichmann
Freiburg, August 8, 2002

Part I: Symposium

The Biology of Complex Organisms –
Creation and Protection of Integrity
Ed. by K. Eichmann
© 2003 Birkhäuser Verlag/Switzerland

Evolution of development

Lewis Wolpert

Royal Free and University College Medical School, University College London, Dept. of Anatomy and Devel. Biology, Gower Street, London W1IE 6BT, UK

Thank you very much, I am flattered to be invited to talk at your celebration – congratulations on your successes. The relationship between evolution and development is a very fashionable field at the moment. My approach is rather different as I focus on how the embryo and multicellularity evolved. I do not know the answers so it is in a sense a just-so story. Rudyard Kipling wrote wonderful just-so stories. There is, for example, a tale of how the camel got its hump and how the elephant got its nose – the crocodile pulled it; my story is rather of that category, it is a just-so story. The supporting evidence is not very strong.

The basic organisation and functions shared by all eukaryotic cells but not prokaryotes, must have been present at least 2 billion years ago, before single-celled eukaryotes diverged. This conservation would include their large size – 1,000 x the volume of the prokaryotic cell – their dynamic membranes capable of endocytosis and exocytosis, their membrane-founded organelles like the nucleus, mitosis and meiosis, sexual reproduction by cell fusion, a cdk/cyclin-based cell cycle, actin- and tubulin-based dynamic cytoskeletons, cilia and flagella, and histone/DNA chromatin complexes. These ancient processes which evolved in the single-celled prokaryotes and early eukaryotes long before metazoa, constitute the core biochemical, genetic, and cell biological processes of metazoa.

These eukaryotic cells were doing very well. Why did they bother to get together? And what had to be invented in addition to what the eukaryotic cell already had to make the embryos? Let me make my position clear. The miracle, and I do not mean it in the religion sense, I mean it in the evolutionary sense, the miracle of the evolution, is the cell. While there are theories involving an RNA world and self-organising, it remains a mystery. Once you had the eukaryotic cell from the point of view of evolution and development it was downhill all the way, very very easy.

Development requires turning genes on and off, cell signalling and transduction and cell motility. The ancestral cells had these. Lower eukaryotes, such as flagellates, slime moulds, ciliates, and yeast cells, have many control mechanisms known from metazoa. Cell differentiation depends on different genes being active in different cells and the cell cycle can be thought of as a developmental programme. There were kinases turning processes on and off, and also genes being turned on-off. There was also signal transduction of stimuli arriving at the cell membrane. Signalling in unicellular eukaryotes was believed to be confined to mating factors in, e.g., ciliates and yeast cells. It is now evident that unicellular eukaryotes depend

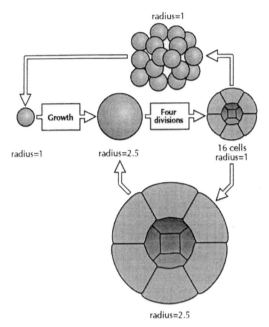

Figure 1. Generation of multicellularity by growth and division.

on extensive signalling systems for their existence. There are also many similarities of the intracellular transduction systems in uni-and multicellular organisms. Eukaryotic cells had motility and chemotaxis. While the slime moulds have nothing to do with the origin of the embryo, they branched from the metazoan line shortly before plants and fungi, they have cell-cell signalling involving several components shared by metazoa, such as cAMP, G-protein linked receptors, a variety of protein kinases, and JAK/STAT transcriptional control. From their unicellular past, early metazoa had a lot to draw upon in the evolution of intercellular signalling.

Single cell organisms have molecular motors and these could provide the forces for morphogenesis. Chemotaxis in the slime mould *Dictyostelium* provides an important model for cytoskeletal organisation and signal transduction and chemotaxis is important in its own right. The chemotactic cell is polarised and polarity is fundamental to many developmental processes. Ligand binding leads to rearrangement of the cytoskeleton-actin polymerisation at the anterior end and results in filopodial extension while myosin at the rear contracts to bring it forward. This illustrates how complex the cytoskeleton already was.

So I want to say my first argument is that eukaryotic cells had everything. We know they did not have collagen or other extracellular matrix molecules , but I am not impressed by that argument as that could come easily. Among the basic components required for development, I can think of virtually nothing that eukaryotic cells did not have which is required for the developmental processes. And so, the real question is why did they bother to get together and what was the adaptive advantage?

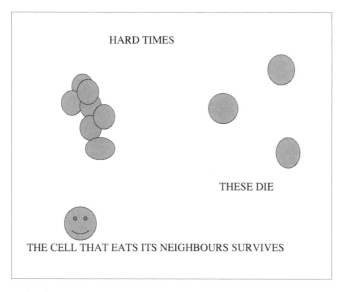

Figure 2. Selection for multicellularity by altruistic cannibalism.

I originally thought that a cell by mistake grew large and then subdivided into quite a large number of cells (Fig. 1). But I could not see the adaptive advantage of that. There are arguments, that being multicellular allows it to have division of labour and specialised functions. But that cannot have been the original selective advantage. So, our argument goes like this. There was mutation in a single cell so that when it divided the cells stuck together. Further division resulted in a loose colony of cells. Now this had an advantage or might have been an advantage if there were other sorts of unicellular predators around. They were more difficult to eat. But what was the real advantage? In hard times when there was no food around and single cells could not survive, some cells in the colony could then eat each other and so survive. If some cells died, then cells could eat them and survive, and that is a major advantage (Fig. 2). It is also the origin of cell death.

There is current evidence to support this idea. If you take planaria or hydra and you starve them, they get smaller, keep their normal form and the way they do is by the cells eating each other. The way the sponge egg develops is by phagocytosing neighbouring cells. In certain fish and annelids what happens at the time of reproduction is that the adult eats an enormous amount of food and devotes all of its body almost to feeding the egg. In fact, muscle cells are actually broken down and phagocytosed, the eggs are laid and the animal dies.

At a later stage in the evolution of this simple multicellular organism it was an advantage to identify the cell which was going to be fed by the others and become large. This is the origin of the egg. The cell might have been under different external conditions, for example near the centre of the colony, such that it would enable it not to die, and therefore to feed on the neighbours. The other advantage of course of having the egg was that it avoided general conflict. While a colony of inde-

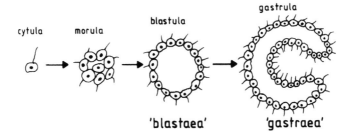

Figure 3. Gastrulation.

pendently reproducing cells could have been successful, mutations in all the individual lineages would have occurred and accumulated. This would have had two severe disadvantages. The first would have been at the level of how cells interacted in the colony. The cells would acquire different genetic constitutions and this would have led to competition rather than cooperation between the lineages. Secondly it would have been difficult for the colonies to lose deleterious mutations or mutations in general, including those reverting to unicellular state. The solution to these problems lay in the evolution of the egg; if the various colonies arose from a few germ-like cells with low mutation rate, then the competition and the mutation problems would both disappear. It is not too difficult to imagine a series of mutations which would have given the inner cells an advantage with respect to eating their neighbours so that in hard times the outer cells died. Our origin, I claim lies in what one might think of altruistic cannibalism.

Haeckel really played a very important role in thinking about evolution and development. He had the idea about ontogeny recapitulating phylogeny which turned out to be incorrect. Ontogeny does not recapitulate phylogeny, the reason why we have something like fish gill slits in our embryonic development is that we partly retain ancestral early embryonic stages. However on one evolutionary change I think he was right, and that is in relation to gastrulation. He has near the bottom of his evolutionary tree what he calls the gastraea which evolves from the simple blastea (Fig. 3). Here there is real evidence that ontogeny really does recapitulate phylogeny. All animals pass through a gastraea – like stage, they gastrulate. Why is gastrulation so similar in all animals? I want to argue that it does actually recapitulate an ancient ancestor.

There is a very simple organism, *Trichoplax*, which is made up of just a single layer of cells and a hollow interior. It is rather like Haeckel's blastea. What is remarkable in *Trichoplax* is that it undergoes a change similar to early gastrulation while feeding. Particles of food or microorganisms that it is going to eat are moved into a digestive chamber.

The basic idea is that a two-layered primitive organism fed on the bottom and it formed an infolding to aid feeding. This basic idea comes from Jaegerstern. In a blastea-like organism, a hollow organism made of a single layer of cells, the feeding was encouraged by currents from cilia.

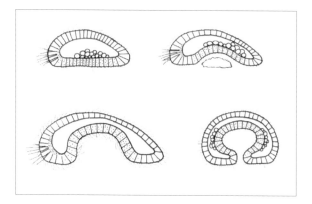

Figure 4. Evolution of the gastrula from feeding.

Living on the bottom it formed an invagination to sweep the food into a primitive gut where the cells are that would engulf the food (Fig. 4). It takes no stretch of the imagination to see that all it had to do is to fuse this infolding with the sheet on the other side and you then have a mouth, a gut, and an anus.

Little had to be invented to reach this stage, everything was there in the cell, and in a way the embryo is really much less complicated than individual cells. The complexity of the developmental biology does not lie in the embryo, but lies in the individual cells. If you look at the signals between cells there are less than 10 grand families of signalling molecules between cells and this is trivial by comparison with what goes on inside cells.

One of the things that need to be thought about in relation to the evolution of the embryo is what of the selection pressures on the embryo itself. Now, I like to liken the embryo to medical students at my university. Medical students can play around, not come to lectures, spend their parents' money, get drunk. Only one thing matters, they have to pass the final exam. It is the same with embryos. They do not have to look for a home, they do not have to mate, and energy expenditure is trivial; the only thing they have to do is to reliably give rise to the adult. About 25% of the cell's ATP goes on keeping sodium out of the cell. So just being alive is expensive. Making one more gene or movement from an evolutionary point is trivial. What matters is reliability. There may be little selection for a more efficient way of, for example, gastrulation. Cnidaria take very different pathways to the planula (Fig. 5). This is a standard way by invagination giving it with two-layers but in yolky egg cell death is the mechanism.

How did cells in the colony evolve different identities and so establish well-defined patterns of cell activities? There are really only two major mechanisms by which cells acquire identity, one is by asymmetric cell division together with cytoplasmic localisation. The other relies on interactions (Fig. 6). An important mechanism in the evolution of the spatial patterning of the embryo could have involved the Baldwin effect which I learnt about from one of my teachers, Conrad Waddington. An environmental stimulus, such as contact with the substratum which

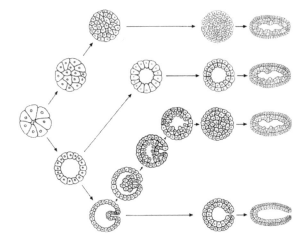

Figure 5. Cnidarian forms of gastrulation.

brings about a change in the cell so that it becomes different from its neighbours, then having established that machinery it becomes intrinsic. For example there are mutations which lead to the thickening of the soles of the feet due to pressure, then this thickening can become genetically determined and occurs only in the feet. I think it is a nice idea, because I think to get thickening autonomously localised to the feet from the very beginning would be much more difficult. And in the same way, one could think of the cell like this, for example, just having contact with a substratum and therefore this region could, for example, secrete a protein and then this could have been used by neighbouring cells, and so perhaps morphogen gradients and positional information could have evolved. Then later on could this become genetically determined. In the same way the cell at the centre of that colony that I

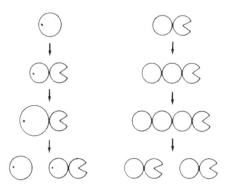

Figure 6. Early development – asymmetry or interaction?

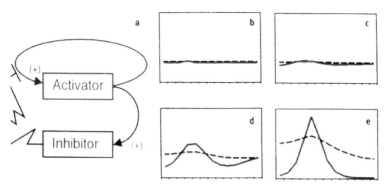

Figure 7. Turing's model of pattern formation.

spoke about at the very beginning, the one that is going to be fed by all the other cells, could make use of the Baldwin effect. It was in the centre of the colony, would have a slightly different environment, and therefore become adaptive and eventually become genetically determined as the egg.

When one is thinking about how patterning evolved it would be very attractive if it used reaction diffusion. The idea of reaction diffusion came from that British genius, Alan Turing who broke your code in the war. He showed that if you had two diffusible molecules, one an activator and one an inhibitor and the activator had a positive feedback, but it also stimulated the inhibitor which inhibited the synthesis of the activator, then with appropriate values of rates this system would self organise (Fig. 7). You could start off with uniform concentrations but it would develop a peak in the activator. And if you increased the length you could get 2 peaks. It is an important possible mechanism of self-organisation. Now, it would be lovely to believe that early patterns in the developing embryo could be based on such a mechanism. Alas, there is not a shred of evidence for it. I do not know of any evidence in the whole of the biological literature which persuades me that the reaction diffusion is actually in operation.

Gene duplication as an important mechanism is so obvious to an audience like you, I am not going to pursue but of course gene duplication was absolutely essential; the virtue of gene duplication is that the cells have now two copies of a gene that is functioning, and so they do not need the other one and it can acquire a new function. And of course, I think that was absolutely fundamental to the evolution of development of the embryos.

The development of early embryos was probably rather messy, they were not very reliable nor canalised to limit the effects of variations in unrelated genes. It did not matter, they had time. We are talking of hundreds of millions of years for things to evolve. They could play around as I told you, so long as some of them passed the exam and gave a good phenotype. Reliability in development has not received the attention that it deserves. Odell's group made a mathematical model to explain an early aspect of patterning in the *Drosophila* embryo in relation to segment polarity genes. What turned out was a surprise, as with their complex network of gene interactions with detailed parameters and rate constants, they could get the same results with quite large variations in the rate constants. Yet if you take isogenic nematode worms you let them grow up under exactly the same conditions. They all die after about 15 days, but there is considerable variation in the time of death. The idea that clones are identical should be treated with suspicion. Examination of the hypothalamus of human identical twins shows that the number of cells in the hypothalamus can vary by as much as 20%. Development is reliable, yes, it passes the exam but sometimes there are variations in the degree of success.

Division of labour could provide advantage once organisms were multicellular. This has been widely claimed to be a crucial step in the evolution of multicellular organisms, though the benefit has never been established. Bell & Koufopanou, for example, suggest that the unexpectedly high rates of increase shown by colonial algae are made possible by the division of labour between somatic and germ cells. For example, if the somatic cells are a source and the germ cells a sink, then there is the possibility that end product inhibition which may act as a negative feedback mechanism for resources could be reduced.

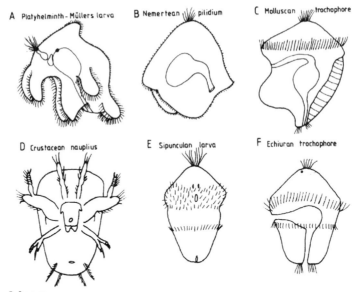

Figure 8. Larvae.

Larval forms are extremely wide spread throughout the animal kingdom and so is metamorphosis, that dramatic change from the larvae into the adult. And the question is how could it have evolved and why do so many larval forms look rather similar (Fig. 8)? It is generally accepted that the larvae of insects were intercalated. The original fly already was a fly-like animal that would develop wings. Only later did they evolve larvae to make use of vegetation and food. However Eric Davidson and his group have argued that in evolution, and this is important from the point of view of evolution of development, the larva is the primitive form and adults came

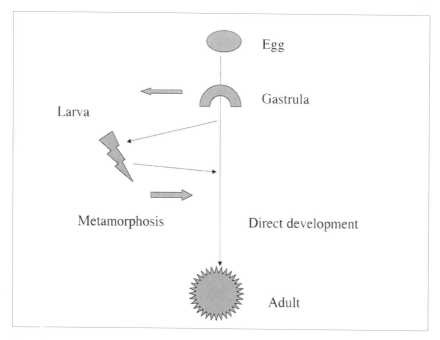

Figure 9. Intercalation of larvae.

later. The argument they make is the following: that the larvae were primitive and then for reasons we do not understand a group of cells called set-aside cells developed. And they are those that in later evolution gave rise to the adult which involved metamorphosis. Therefore, the larval form was the original organism and metamorphosis evolves later.

I wish to argue this is totally and utterly impossible. Just think of the frog. When a frog begins to develop a larvae is made, the tadpole, and then you get meta-morphosis. I cannot imagine any of you wishing to argue that the tadpole was what evolved into the frog. What clearly happened with the frog was that an early talebud stage in normal development became a bit motile, it swam away a little bit. What is the function of a larva? Larvae only have two functions, one is dispersal. The last thing you want to be if you want to succeed is to be with all your siblings. When the eggs are laid, you want to get away. The other evolved later was the ability to feed and therefore to get further away and to be able to get bigger. And do not think any rational person really argues about the tadpole being primitive and evolving into the frog.

For these reasons the evolution of larva and metamorphosis must be due to intercalation of the larval stage into a directly developing animal (Fig. 9). A stage of embryonic development or early growth becomes modified to form the larva and that metamorphosis is essentially a return to the original direct developmental program. It is not possible to image a scenario in which set-aside cells in a larval-like form could evolve to give an adult whose form is different to that of the larva.

It is only by modification of the direct development that metamorphosis becomes possible.

One of the points that Davidson rightly makes is that the larvae of very different groups are extremely similar. This is a quip from one of their papers, in which it says 'It seems to us the epitome of hand waving to accept that the larvae of distinct sets of phyla could be due to parallel evolution'. I intend to wave my hands because that is precisely what it is. And let me now persuade you that that is the case.

In the metamorphosis of the sea urchin larva into the adult quite a lot (by no means all) of the adult comes from a small group of cells on the left hand side of the larva, and these are what Davidson calls set-aside cells. I will offer a bottle of champagne to anybody who can give a plausible scenario whereby in evolution first of all you could have selection for set-aside cells; but more important, given a small group of cells like this, that could evolve in the future metamorphosis to give rise to a sea urchin. I claim it is totally and utterly impossible. I wish to argue the complete contrary, namely that all larvae are intercalated into direct development just as in the case of the insect and the frog. What metamorphosis is about is that during direct development an early form shortly after gastrulation becomes the larva, able to disperse. Metamorphosis is to bring it back onto the direct developing programme and to develop into the adult. So these are two very strong counter views.

The reason why larvae look alike is because of gastrulation. They all developed from a gastrula and all invertebrate gastrulae are rather similar. Here is a recent paper on the protostome and the deuterostoma and the gastrulae in both; the way the mouth forms is fundamentally different. In the sea urchin for example you get the invagination, this structure will become the anus, this structure will become the mouth. In the proteostomes there is a complicated process where this structure becomes divided and the mouth and the anus form from the same region, but ultimately end up very similar as larvae and closely related to the gastrula. And so, the hand waving is not that it is surprising that all larvae look alike, it is inevitable. Because that is how they evolve from the gastrulae which are so similar in very, very many forms. Incidentally, once larvae had evolved, the evolution of metamorphosis to give the adult form is quite complicated, and it is hormonally based. All the current evidence is that the hormones that are used were those that were related to those already there, because they were for the control of growth in the adult.

One of the problems in evolution of development is how the complexity of signal transduction can be understood. This can be illustrated by a Rube Goldberg cartoon and it is my model for signal transduction whose design I do not understand. The signal is rain, and the end result is to ensure that a person can smoke his cigar, and so an umbrella goes up. So the rain falls on a prune, which moves a lever, which lights a lighter, which lights a candle, which boils this kettle, whose steam blows a whistle which frightens a monkey. This monkey jumps onto a swing which cuts a cord holding a balloon which goes up, and releases the door of a cage so birds fly out and pull the umbrella up. I think that is a simplified version of signal transduction as there are many monkeys.

I think it is wonderful, but the question I am interested in is how could this have evolved? How could all that signal, and the only clue that I can offer to you is from hearing a talk by Adrian Thompson, who worked at the University of Sussex. There is an interesting group at the University of Sussex, where biologists and engineers have got together and the engineers are making use of biology to apply to engineering, one of the things which they have. Adrian uses genetic algorithms to design a circuit, he is an electronic engineer and he wants to design a circuit in which he says, Go, light goes on, and he says Stop, light goes off.

He sets up on his computer little circuits almost randomly connected to the unit. Then he looks, he tests them to see if any have some of the properties that he is looking for. And then he mutates it and repeats the process. He goes through around 600-800 iterations. In each case he is simply selecting the circuit that is getting closer and closer to what he wants. At the end, he has a circuit that works. He then looks at the circuit and has not the foggiest notion how it works, it takes him 3 to 4 months to find out how that circuit actually works. In some cases it has actually invented a clock. And the point that I am making is that when we come to look at signal transduction and many of the other pathways we are looking at them as we would have designed them. Evolution is not like that. All that matters is that it works. I think the experience of these genetic algorithms here is a very important way to begin to try and understand and get some insights into the evolution of development.

Thank you for listening.

The Biology of Complex Organisms –
Creation and Protection of Integrity
Ed. by K. Eichmann
© 2003 Birkhäuser Verlag/Switzerland

The cloning of mammals: what are the problems?

Rudolph Jaenisch

Whitehead Institute for Biomedical Research, 9 Cambridge Center, Cambridge, MA 02142, USA

I want to talk about a topic which is of public interest, the cloning of mammals. When you look at the media, for example the New York Times Magazin´s cover a couple of months ago, what emerges is the issue of the mad scientist, of out of space sects which believe that life came to earth by cloning, and so on. I think this is a ludicrous debate and I am not going to go further into that. There are rather serious and interesting issues behind this and these issues have been raised already half a century ago, in the seminal experiments with frogs done by Briggs and King, Gurdon, DiBernardino and by others. The principal questions which were posed then were: Does differentiation involve loss of nuclear potency, and is there nuclear differentiation? Can a nucleus of a terminally differentiated cell be reprogrammed to participate in development of an animal and the differentiation of all lineages. These are essentially the same questions we ask now, in one way or the other, and I will discuss these questions. In the second part of my talk I will actually bridge the topics of this institute. I will talk about immunology, including B and T cells.

Let me come to history. In the 50ies and 60ies the people I mentioned above took donor nuclei from either early frog embryos or from later embryos, tadpoles, from gut cells for example, implanted them into enucleated oocytes and asked the question, can they direct development to animals like tadpoles or adult frogs. And very rarely, Gurdon indeed got frogs. However, success was so rare that the origin of the donor cell was never really certain. I will come back to this problem. I think this is a major problem also in present day cloning experiments. The interpretation of the results was yes, the nucleus can retain totipotency but this appears to diminish with development. This was not a really clear conclusion. In the early 80ies, a very important experiment was done by Davor Solter by transferring nuclei from very early cleavage stage embryos into enucleated zygotes. He observed that all clones died within a few divisions. I think that these were very carefully controled experiments. It was a very important issue at this time and it suggested in a way that mammals could not be cloned. By another approach, using the oocyte as a recipient, it was clearly shown later that mammals could be cloned, for example sheep (Dolly) and then a number of other species that have been cloned in succession. What are the questions here? I think nuclear cloning is a really important tool for science. A number of different questions concern the epigenetic state in particular, in which I have been interested for a long time. We have a number of cell types in the body, we have embryonic stem cells, we have adult somatic stem cells, most of which are elusive, and we have terminally differentiated cells. One of the issues this technique will allow us to address is, is the potency to differentiate into different lineages, which is inherited in these cells, does this correlate with the efficiency to serve as a donor for nuclear transplantation? Is there nuclear programming? Does the efficiency in nuclear transfer depend on the epigenetic state of the genome? In other words, is it a general tool to look at the epigenetic state of the genome in general? These are some issues which are quite interesting and will be followed, I think, using this technology.

Let me remind you that reprogramming occurs in normal development. It is a very normal and very important event. The epigenetic state is reset at two stages in the developing embryo or before. It is reset during gametogenesis, involving the whole genome essentially, and prozygotically, involving the choice of the active and the inactive x chromosome, and the adjustment of telomer length. Let me just point out what I mean with this. If we look at the primordial germ cell, then we have much evidence that the genome in the primordial germ cell has been really – if you want – reset. It is certainly hardly methylated, unmethylated probably. The chromatin is not really analyzed in these cells, which are very difficult to isolate. Presumably it is also in a reset state. During the complicated process of gametogenesis the mature gametes, an egg and a sperm, will finally be produced. One of the results is that these cells are competent, their genome is competent to express the genes that are needed for early development after the two gametes come together at fertilization. I think that is a very important step. Clearly other things occur during gametogenesis, meiosis occurs, recombination occurs, but I think one of the more important events which occurs in gametogenesis assures that development can take place when the two gametes come together. Sets of early genes are activated rather efficiently and faithfully. At a later stage only x chromosome inactivation occurs and telomer adjustment, events which do not have to occur during gametogenesis. The final outcome of this is that adult tissues are generated which do not express the early genes anymore. Instead, they express specific sets of adult genes.

Of course, there is another population of cells in presumably most tissues which are called somatic stem cells. They probably in some way resemble, I think that is important, the early embryonic cells, for example, embryonic stem cells. This will be important for what I am going to have to tell you. These somatic stem cells, they are basically undefined. This is the way by which resetting of the genome occurs in normal development. In cloning, one takes one of these adult cells, which express a set of adult tissue-specific genes, and after transfer of the donor nucleus into the recipient egg the genome has to be reprogrammed. There are three possible outcomes. Either there is no reprogramming and the embryos will fail very early. Or there is partial reprogramming, so the embryos will develop abnormally, or there is full reprogramming and these embryos will be normal. Now, we know from all five species cloned so far, the latter is the exception. Most clones die at various stages of development. If they are born, they will be possibly abnormal, as we see in the majority of cases. One problem is that reprogramming has to occur very rapidly, probably within hours after the transfer, because the egg has to divide. There is not much time, and it has to occur in a cellular context very different from that in gametogenesis. So, let me make a few predictions. There should be little or no problems for reprogramming in events that occur after fertilization. This is x-inactivation and telomer length adjustment and indeed we have shown that the x chromosome which is chosen to be active or not active in the donor cell is perfectly reprogrammed and this is done by random choice. No problem. Telomer length adjustment has been more controversial. People thought because Dolly is older than her biological age, the 6 years her donor was in addition to her own age, her telomers should be shorter. The group from Edinborough indeed reported they are a little bit shorter. I do not believe that this is the final answer because in cloned cows you find they are either normal or they are even longer. I think this is predicted, because telomerase is expressed in post-implantation embryos and there is no reason to assume that the telomers should be shorter . These are not the real problems. The problems are really in the re-programming events that have to normally occur prior to fertilization, concerning essentially the whole genome, imprinted and non-imprinted genes. The real problem of nuclear cloning is that the donor cells expressed tissue-specific genes, in the case of Dolly probably genes for milk production, but did not express genes which are needed for embryonic development, like the Oct-4 type of genes for example. The transplanted nucleus must probably activate those genes in order to succeed in directing development, and possibly has to inactivate the tissue-specific genes, though we do not know that.

So, let me briefly show you how cloning is done in mice. In a short movie which a student of mine took, I show you a method which was devised by the Hawaii group by Terry Wakajama, with whom we collaborated very closely. For mice it turns out that they are very different from all the other organisms which have been cloned. In other species there is a fusion step which actually makes cloning much easier, whereas in mice the nucleus has to be injected. Here is the oocyte to be enucleated using an enucleation pipette. In the Hawaii method it is important to use a very blunt pipette, it is inserted through the zona pellucida and through the cytoplasmic membrane later by the piezo drill element. This makes it really go in very very smoothly – as you can see – through the zona pellucida. You do not butcher the embryos the way you do with the conventional way of injecting nuclei or taking something out of the cell. It is a very smooth way to getting through, you can see the sucking out, you suck out the spindle, it is very difficult to see the spindle. The preparation of the donor nucleus is done mechanically with this small bore pipette, you suck the cell up and down to break the cytoplasmic membrane and leave the nucleus intact. So, you suck up the nucleus, several nuclei into the nuclear transfer pipette and you free it more or less from cytoplasm. This is very different from sheep cloning or cow cloning where you subject the enucleated oocyte and the nuclear donor cell to electric fusion. You put the cell under the zona, give an electrical shock and just fuse the whole cell into the recipient egg. This is probably much less damaging to the cell. Now comes the transfer of the nuclei into the eggs and again you have to go through the zona and then also through the cytoplasmic membrane, you do this with a piezo drill element which gives you this very fast vibration of the blunt pipette. Now you will see how the nucleus is expelled and will go very deeply into the cytoplasm of the oocyte – you brake the cytoplasmic

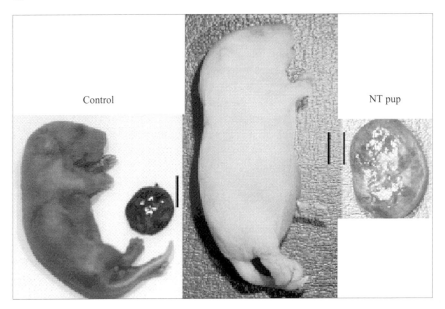

Figure 1. Cloned mice often display severe overgrowth.

membrane and deposit the nucleus into the oocyte, thus generating a diploid cell again. The technique is not easy, it is very high power, and I think many laboratories have tried it and in our experience it needs a lot of practice. But once you can do it you do 150 or so of these transplantations per hour, get it done in the morning and then have some animals surviving to later stages.

In general, nuclear cloning is very inefficient, most clones die soon after implantation, very few survive to birth. This is true for all species. There may be serious abnormalities, they may die pre-natally or post-natally. And very few become apparently normal adults. So, again going back in a way to Gurdon and to Briggs and King, one of the questions is, does survival of a nuclear clone depend on the donor cell type. I think this is the very key question. Here I summarize many experiments, particularly from Wakajama, comparing his results using as donor cells somatic cells (cumulus cells, fibroblasts, Sertoli cells), and ES cells. You can see, survival to adults is really extremely low, it is in the order of few percent or less. However, we found and they also found that ES cells – when used as nuclear donor – are much more efficient in giving rise to adults, probably an order or so of magnitude higher. We surely argue that the ES cell nucleus, being an embryonic nucleus, might be easier to reprogram. And this will be very important for what I am going to tell you later. Nevertheless, whatever you use as donor, a cumulus cell or an ES cell, you end up with the large-offspring-syndrome. In all species the most common problem seen in cloned newborns is fetal overgrowth, in addition to defect of placenta, respiratory distress, abnormal circulatory systems, and all of those might be the immediate cause of death. Once they are born, if they cannot inflate the lung they are dead. That happens very often. Fetal overgrowth and defect of placenta is very characteristic. You see this even in the cloned animals that survive, probably all animals, even the 20 so-called healthy cows which have been published a week or two ago in Science, they certainly went through that problem. The fetalmaternal relationship was not normal. Although there is no normal placenta, they might recover to develop to apparently healthy adults, but often they turn out to have an abnormal immune system, there may be an increased number of infections. They may also have skeletal abnormality, they may have any abnormality you want. So, it is a very stochastic problem which you see frequently. Let me show you this one example of the large-offspring-syndrome in mice (Fig. 1). This is a normal newborn pup with its placenta, this is one of these large newborns which was 4 times the size of a normal one, whereas the placenta was actually 7 times normal size. This is an extreme case and this animal did not survive, it died soon after delivery. You see this highly abnormal umbilical cord which reveals problems in the circulatory system.

When you see such animals, of course, you think about the class of genes that are imprinted, because imprinted genes are known to be important for fetal growth regulation. So, we looked at imprinted genes. Let me just give you a brief introduction to imprinting for those who are not familiar with it. This is based upon nuclear transfer experiments done in the beginning of the 80ies and particularly first by Davor Solter. A normal bi-parental embryo has both a maternal and a paternal genome. Uni-parental embryos were generated by nuclear transfer, by replacing the male pro-nucleus by the female pro-nucleus to generate gyno-genomes or replacing the female pro-nucleus with the male pro-nucleus to generate andro-genomes. These uni-parental animals fail in a very characteristic way. The gyno-genomes have placental defects, no muscles, good brain development, the andro-genomes have large placentas, extensive muscles and no brain. So, what came out of those experiments was that the two genomes have not the same function, they have complementary functions. So normal animals need both, actually. This is of course due to imprinted genes, and I give you one example of imprinted genes which is the best understood: Igf-2, a fetal growth factor which is paternally controlled, and H19,IGF-2r which are maternally controlled and counteract the growth promoting effect of Igf-2. The size of the offspring critically depends on the balance between these two classes of genes. And this is probably true for many imprinted genes, although we do not know the details for others pairs as well as for those. So, we looked at the expression of these imprinted genes. The expression of the imprinted genes in the normal newborn, either in placenta or in organs, is pretty constant. But if for example you look at H19 in clones, it does not matter where they came from, H19 varies widely between values which are higher or as high as in controls to total extinguishing. You can see this to a similar extent, not as extreme, for Igf-2. Some clones really overexpress this growth factor, others do not. This is true for the imprinted genes and we looked at many of those genes. I am not going to bother you with any of these analyses at this point. I just want to summarize the results. The cloned mice showed pronounced dysregulation of imprinted genes. None of 50 cloned pups expressed all of 6 imprinted genes that we have tested in detail. None of the other genes was dysregulated. Dysregulation of a given gene did not correlate with normal or abnormal expression of any other. They were all independently dysregulated, it is stochastic. And most clones of course showed the large-offspring-syndrome. Now, it is important to mention here where this variation comes from. There was first of all no correlation of gene expression with fetal growth abnormalities, we were not surprised to see this because there are not only six imprinted genes, there are probably 200. So when we look at a correlation with 6, of course, we would not see this. We did not expect it either. In addition we know now that imprinting is not the only answer for the large-offspring-syndrome. If you look in pups that are derived from cumulus cells that are never cultured and which actually have pretty intact imprinting, they still have the large-offspring-syndrome, and we know there are non-imprinted genes involved.

In summary, I think the problems one can see in cloned animals – particularly when you think about ES cell-derived animals – are due to pre-existing variation in the ES cell population as well as to faulty reprogramming. Let me come to the pre-existing variation, and I think that is important. I am not going to go into any experimental details but what we concluded was that ES cells display just a surprising epigenetic instability. So when we looked even at ES cells which were from a subclone of ES cells, sibling cells so to speak, they varied in the H19 expression, in the H19 methylation and so on and so forth. This was surprising to us. The variable expression of imprinted genes in the ES cell-derived nuclear transfer pups was determined by a combination of two factors, the pre-existing variation in donor cells and the epigenetic instability in the ES cells. We believe this is probably due to culture, any cell that has been cultured loses imprints probably irreversibly, and due to faulty reprogramming. The issue of course which is raised by this is the possible instability in human ES cells, which might be important for the use in transplantation. I think that needs to be analyzed. However, we do not believe that it will be a problem. The instability is clear for most ES cells but I think it does not play a role when you use these cells for transplantation. So one conclusion from all of this was that cloned embryos may develop to birth despite the wide-spread dysregulation of imprinted genes and of non-imprinted genes. The other extreme possibility would be that a single important gene is dysregulated, but this is not likely to cause death in mammals. Mammalian development is quite tolerant to abnormal regulation of even multiple genes. Up to a certain threshold, of course, if you do not activate a key embryonic gene you are dead, but you can dysregulate quite a lot of genes and you are still alive. So, the problem of lethality or abnormal phenotype may arise early when key embryonic genes are dysregulated or later when later acting genes are dysregulated. I think the important conclusion is that apparently normal adult clones may have abnormal regulation of many genes but this is just too subtle to result in an obvious phenotype. And this applies also to the experiment recently published in Science where they reported just the physical examination of 20 cloned cows. Of course, they found the liver has the right size, of course they found that the serum values were normal, otherwise they would be not adults. They conclude they are normal, but I think this is a very premature and unacceptable conclusion. They have to look in a much more sophisticated way at these cows. But anyway, that is where these public discussions of course become important.

Let me come to the second part and a different issue. I think the issue again goes back to the old frog experiments: What is the nature of the donor cell for the rare successfully transplanted nucleus? Because cloning is so inefficient this becomes a very important question. Let me recall the facts. The facts are that the ES cells are much more efficient nuclear donor cells than somatic cells, I told you this. Somatic stem cells certainly have similarities to ES cells. They have a high potential to differentiate into different cells. So, the epigenetic state might be similar in some way to ES cells, though we do not know that. The key question asked by Gurdon, and by others in response to Gurdon´s experimentation, was: Are clones derived from differentiated cells, or are they rather derived from rare somatic stem cells which are present in the donor population? This issue was raised immediately after Dolly was generated. People asked: is Dolly really derived from a terminally differentiated cell? Of course, it was really impossible to answer that question. The possibility that it was a stem cell nucleus could not be excluded because there is just no unambiguous marker that identifies the actual donor nucleus that had generated Dolly or any other clone. So, a student of my laboratory, Konrad Hochedlinger, suggested that the way to do this would be to generate a monoclonal mouse derived by nuclear transfer from a terminally differentiated lymphoid donor cell. In this case the rearrangement of the immunoglobulin genes or the T cell receptor genes would be an unambiguous and stable marker for terminal differentiation. This experiment had been tried before, for 4 years by very skilled people, and it had failed, the embryos mostly died very early. In some cases they got animals, but they had no clue what was the donor. This was exactly the old problem. Therefore, Konrad adopted a three-step approach. In a first step, he transferred nuclei from a peripheral lymph node into oocytes and derived blastocysts, cloned blastocysts. These blastocysts, instead of getting injected directly into the fostermother, were put into a petri dish to generate the ES cells. And then the ES cells could be used at leasure to generate mice. In this way he eliminated one of the bottle necks in the normal cloning. Everything has to be in place at the time you put the blastocysts back into the foster mother, and if something goes wrong the experiment cannot be repeated. His experiment was now the following. He isolated peripheral lymph node cells, took a nucleus from those, transferred this to an enucleated egg and got blastocysts. The blastocysts were then explanted into the petri dish to get an ES cell line and the ES cells then were used to generate mice by various procedures. I briefly tell you what was found.

So, it turns out this approach is exceedingly inefficient. We injected 1000 oocytes, we got 40 blastocysts and 2 ES lines. This is in contrast to similar experiments done by Makajama, where they used fibroblasts or cumulus cells as donors. They got out 60–70% blastocysts and probably an order of magnitude more ES cells. With respect to the fertilized embryo – here probably they were a couple of hundred times more efficient in getting ES cells. So clearly using lymphoid cells was very inefficient. Moreover, when we got only these 2 ES cells, we suspected they would not be from B or T cells, because a lymph node has maybe 95 % B and T cells and 5% non-lymphoid cells such as macrophages, dendritic cells, stem cells, anything. So, our prediction was, it will be from those. Of course, he had ES cells, so he can look at

DNA and he looked at DNA; what he found was rather satisfying. He probed the DNA with a probe for the immunoglobulin heavy chain locus and he found that in a wild-type ES cell, of course, it is in germ line configuration. In contrast, LN1, the first ES cell derived from a lymph node cell, had apparently rearranged both genes of the heavy chain, and also one of the two κ-chain genes in this cell was rearranged, whereas the other was normal. The other ES cell, LN2, had rearranged both the T cell receptor α alleles and T cell receptor β alleles. So it was an unambiguous proof that LN1 was derived from a mature resting B cell which was waiting in the lymph node for its antigen to come by and LN2 was derived from a mature T cell which had rearranged both receptor genes.

The question now was what is the potency of these ES cells derived from peripheral lymph node donor cells, can they generate normal mice with rearranged immunoglobulin or T cell receptor genes in all tissues. He used two methods to generate mice, the one involves tetraploid complementation, it is a method by Andor Snuch. One generates tetraploid blastocysts by fusion of a two cell embryo and these tetraploid blastocysts cannot contribute to the embryo itself, only to the placenta. Embryonic stem cells of course cannot convert into the placenta, so those two complement each other. The ES cells when injected into the tetraploid blastocyst form a completely ES-derived mouse, not a chimera. It is a great method in general and it works quite well with a few modifications from Andor Snuch´s protocol. When he did this with the 2 ES cells, the mouse embryos were totally derived from the ES cells. It turns out LN1 was extremely efficient, from 159 injected blastocysts we got 16 pups, only 3 of those died, so we got 14 animals and they were fertile and healthy.

LN2 was much less efficient, 100 injected blastocysts gave us 1 pup which died after birth with large-offspring-syndrome. So clearly, this cell was totipotent, in essence it could generate all tissues of the body, but it had some other alterations which we see very often and which interfere with the postnatal survival. However, when we injected LN2 into diploid blastocysts from the Rag2 mutant mouse which cannot have their own immune system, then we got actually three surviving chimeras. So, clearly this cell is not as able to promote postnatal survival but it is also totipotent.

So, what is now the status of the DNA of these mice? There was no germline configuration of the immunoglobulin genes, there was only somatic rearrangement, and it is present in all organs, liver, spleen, lung, heart and brain of these clone mice. I should tell you that the donor cell was from a B6xDBA2F1 mouse. B6 only expresses the B allele of the heavy chain genes in peripheral lymphocytes, whereas DBA expresses only the A allele, F1 expresses both. So we asked the question which immunoglobulin allele is the one which is expressed in the cloned animals. The clone only expresses the A allele, so we know this is the productively rearranged allele. And we also know the variable region gene which was used, by PCR it was identified as VH22. In a normal wild-type spleen there is almost no signal on a Northern blot but these clone mice have a very strong signal, this is the rearranged allele which is being used.

In the T cell mice, the only clone pup we had, again there was no germline configuration of T cell receptor genes in contrast to a control mouse. TCRα genes were all rearranged. In the RAG chimeras that have been made the immune system is essentially derived from these ES cells. The chimera has normal numbers of CD8 and CD4 cells. And we know also which Vβ is used, it is Vβ14 , as we know by sequencing. A control mouse has very little expression of Vβ14 as one would expect but in these chimeras the great majority of peripheral T cells do express this particular variable region.

The monoclonal mice have rearranged immunoglobulin or T cell receptor genes in all tissues, there is monoallelic expression of the allele that was productively rearranged in the donor B and T cell. We can show that. I think we can firmly conclude that. And there seem to be no rearrangement and expression of other immunoglobulin or TCR genes. I think this is probably very much in agreement with 20 years of work – as we heard this morning – that rearranged transgenes will inhibit the endogenous genes to be rearranged, although there might be mutations in these genes. Imagine, these animals are fertile.

Monoclonal mice can be generated but this is a very inefficient process. Just to recount, from thousand transplanted oocytes we got only 40 blastocysts and we altogether got only 2 ES lines. This is much less than what you get from other situations. So the reprogramming of the B and T cell nucleus may be so inefficient that direct derivation of cloned animals may just not be feasible except by this three-step indirect approach we did. So one has to do these tricks. The conclusion nevertheless is, nuclei from a terminally differentiated cell can be reprogrammed and are able to generate normal cloned animals. Of course, an unresolved question still remains: are clones derived in the normal approaches from terminally differentiated cells as assumed or in fact from somatic stem cells? I think our experiment does not answer this, I think it can be said that in the hemopoetic system the terminally differentiated cells are so inefficient that if you ever get clones they will not be coming from B and T cells unless you do these tricks.

So, I think this raises an interesting question about epigenetic programming of differentiated cells vs somatic stem cells. Clearly we have done the more complicated experiment, namely, showing that indeed B and T cell clonings were inefficient. The next question now is, is the efficiency higher for nuclei of pluripotent or for committed hemopoietic stem cells? I think our result is important, because now we can compare cloning efficiencies in one lineage. Of course we are also interested in neural stem cells, for example, in post-mitotic neurons. I think especially the last point is a very interesting one. Can we clone post-mitotic mature neurons? The old question behind this is: Does brain development or memory involve genetic alterations as opposed to only epigenetic alterations? Let me remind you: the olfactory system is a very intriguing system where from thousand or two-thousand olfactory receptor genes one is chosen in a given neuron and only one allele. It is not clear how this occurs. If we were able to clone neurons which had made that choice, I think we would get very clean results. Either it was epigenetic, the resulting clone will have less than 1 % of this particular receptor expressed. If it was genetic all of them will, so it is a very clear prediction. Another question: Can one get post-mitotic neurons back into cycle? Of course, very different from B and T cells which can go normally back into cycle whereas neurons may not. So, I think this is in principle a way to address these questions.

Let me come to some of the issues which were raised before, mainly because of this public debate. Let me just summarize what I think are clear lessons from animal cloning. I think, the most important was that even clones that survive to birth have often serious abnormalities and die later and they show widespread epigenetic dysregulation. And this is not only true for clones derived from embryonic stem cells as of course some of these cloning activists always criticize. Apparently, as I said before healthy clones may have some defects, which are not severe enough to cause apparent phenotypes. For example, we do not know about the brains of these cows. I heard that the ACT group, who did also a human cloning now, argued that the social interactions between their clone cows are totally normal. I would argue it is very superficial to look at how cows socially interact. We do not have the ways to look at this in any reasonable form in animals, certainly not in subprimate animals. As I said before I think completely normal clones may be the exception, but these cloning activists really make two main arguments, which I want to bring forward. What you always hear is that IVF 30 years ago was exactly at he same stage as is cloning today. It is inefficient, there is just a need to improve the technique. Now this is a totally false argument. IVF is wholly technical, once the doctors had learnt how to culture the embryos and do the fertilization it became very efficient, in 30% or whatever. So, lots of people are now generated in this way, the problem was clearly solely technical. Cloning is both technical and biological and there is just no progress whatsoever over the last 4 years to improve anything. I think we have just the stage where we realize how big the problem is. To argue now we can solve this, we just have to practice, is nonsense. Then, I think, a particularly wrong argument is, is it possible to pre-screen for abnormal embryos and plant only normal ones. The question is: what is normal, and particularly, can you measure this. They say – we look in clone pre-implantation embryos for the key genes, like Igf2R, is it normally expressed or not. This is total nonsense, there is no key gene. Igf2R, for example, did not show up in our screen in mice, it does not play a role in clones of mice whereas others do. There are at least 200 genes potentially dysregulated in a clone placenta. Then of course, they have heard of expression profiling, right, and they say, we do expression profiling. Again this is wrong, because you cannot expression profile for genes which are not active at the pre-maturation stage. Then they argue that routine pre-natal diagnosis detects genetic abnormalities, chromosomal abnormalities, or known gene defects, very efficiently, and can be used to pre-screen. But the defects in clones, the ones we are concerned about are of course epigenetic, they have something to do with reprogramming. Epigenetic abnormalities just cannot be detected by any technique we have. Even not in single cells in a petri dish, not to talk about an embryo. And of course, expression screening, as I said before, is only possible for genes that are expressed at a stage you analyze and not at a future stage. Again that is nonsense. I think there is just no way to predict that a given clone will develop to be normal or abnormal.

So, let me finish by mentioning some of the misconceptions of this whole debate. I think this is pretty important. People mix up productive with therapeutic cloning and this largely due to the total ignorance of particularly those people who make the decisions. They just do not understand any of this. I think the word "cloning" is particularly bad because it provokes these emotional reactions. It is like the term nuclear. For example nuclear imaging, right, was a technique which just was terrible. Nobody wanted to come into contact with nuclear imaging. Once they dropped the word nuclear it is now used very successfully in the clinic. Therapeutic cloning has the same problem due to the term cloning. It causes emotions against what could be the consequences. And this may impede really quite important potential applications for regenerative medicine. I do not want to go into this. And certainly we are just in the assay phase. Just about to go into some legislation which probably will be unfortunate. So, I think I want to stop here. Thank you.

The Biology of Complex Organisms –
Creation and Protection of Integrity
Ed. by K. Eichmann
© 2003 Birkhäuser Verlag/Switzerland

Size control and timing in development

Martin Raff

MRC Lab. for Molecular Cell, University College London, Gower Street, London WC1E 6BT, UK

It is a great pleasure and privilege to participate in this celebration. I feel close to the areas of science that this institute is involved in, as I was an immunologist in my previous life and am a developmental neurobiologist in my present life.

The problem that I want to talk about is a fundamental one in development. It concerns how the size of an organism or one of its organs is controlled. How is it that we grow to be larger than mice? How is it that our arms grow to be the same size, and so on? The embarrassing fact is that we have no idea. The reason is that size control has not been a problem that has interested developmental biologists in recent years, which is very surprising, given its importance. We know that the major determinant of size is total cell mass, which depends on cell size and cell number. I will talk first about cell-size control and then about cell-number control. I will use as examples two types of myelinating glial cells – Schwann cells, to talk about cell-size control, and oligodendrocytes, to talk about cell-number control.

Cell-size control

One aspect of the cell-size problem is that two cell types from a primate – a lymphocyte and a retinal neuron, may differ greatly in size, we have no idea why. No doubt, the difference partly reflects the different genes expressed. But which genes are the relevant ones? And how much of the difference reflects differences inside the cells, and how much reflects differences outside the cells, such as the extracellular signals that stimulate the cells to grow?

The specific cell-size problem that I want to address is how cell growth is coordinated with cell division so that, when a cell divides, it divides at an appropriate size. The fact is that cell growth – that is, cell enlargement – is as fundamental to life as cell division. No organism can grow unless cells grow. If cells simply divide, they will get smaller and smaller and disappear. Yet, there are hundreds of laboratories working on cell division but only a handful working on cell growth. This neglect of cell growth makes no sense, as growth is as easy to study as division. One reason for the neglect may be that there is not a term for the process. Most biologists use the term cell growth to mean cell proliferation rather than cell enlargement. The term "growth factor" is used to describe extracellular signals that stimulate cell growth, cell differentiation, cell proliferation, or cell survival.

So, how is cell growth coordinated with cell-cycle progression to ensure that proliferating cells maintain their size (Fig. 1)? There are two main ways that this is

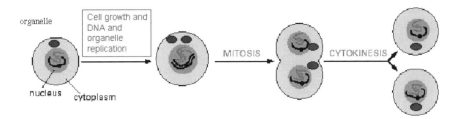

Figure 1. How does a proliferating cell co-ordinate its growth with cell-cycle progression.

thought to be achieved, suggested largely from studies in yeasts. One way is that cell growth may be limiting for cell-cycle progression. The other way (which is an extreme form of the first) is that cells may have cell-size checkpoints in G1 and G2 phases of the cell cycle, where the cycle control system pauses to be sure that the cell is large enough before progressing into the next phase. How cells might assess their size is a mystery. I will talk about experiments done by a very talented graduate student, Ian Conlan, that challenges both of these views of how cell growth and cell-cycle progression are coordinated.

He studies Schwann cells purified from the sciatic nerve of newborn rats. He studies cell size by measuring cell volume in a Coulter counter. He analyzes cell growth independent of cell-cycle progression by blocking the cell cycle in S phase with aphidicolin, which blocks DNA synthesis by inhibiting DNA polymerase α. He finds that the average cell volume in Schwann cells proliferating in serum is smaller than that of cells arrested in S with aphidicolin for 24 hours. Thus, as is known from studies in yeast, cells do not need to progress through the cell cycle to grow; when cell-cycle progression is blocked, cells get larger and larger. To be sure that the Coulter counter measurement of cell volume reflects dry cell mass, which is what Ian is actually interested in, he has collaborated with Graham Dunn at King's College, London, who uses density interference microscopy to study the dry mass of cells. He obtains the same result: proliferating cells have less dry mass on average than cells arrested for 24 hours in aphidicolin. To determine if cell growth depends on extracellular signals, he arrests the cells in aphidicolin and removes the serum and growth factors. As one might expect, the cells do not grow, indicating that cell growth depends on extracellular signals – presumably growth factors present in the serum.

Having set up the system, Ian studied two so-called growth factors that have been shown both *in vivo* and *in vitro* to promote Schwann cell proliferation – insulin-like growth factor 1 (IGF-1) and glial growth factor 1 (GGF-1). In aphidicolin and the absence of serum, IGF-1 stimulates the cells to grow whereas GGF does not. So, IGF-1 is a true growth factor for Schwann cells, while GGF is not. Not only can GGF not stimulate the cells to grow on its own, but when added to saturating amounts of IGF-1, it does not further increase cell growth.

Although GGF does not stimulate Schwann cells to grow, it is a potent mitogen for Schwann cells, in that it can stimulate them to progress through the cell cycle. It induces cells that have been arrested by removal of serum, for example, to incorporate BrdU into DNA. IGF-1 has little ability on its own to do this, but there is an enormous synergy when GGF and IGF-1 are added together.

How is it that these two signal proteins that act on the same class of receptors, receptor tyrosine kinases, have such different effects on Schwann cells? Ian has looked at two intracellular signalling pathways – the MAP kinase pathway, assessed by the expression of tyrosine-phosphorylated Erk1 and Erk2, and the P13 kinase/Akt pathway, as assessed by the expression of tyrosine-phosphorylated Akt. He finds that IGF-1 induces a sustained activation of Akt but only a transient activation of the MAP kinase pathway, whereas GGF does the opposite. Thus, it is the kinetics of activation along these pathways that seems to account, at least partly, for the different effects of IGF-1 and GGF on Schwann cells.

It is known that mitogens work, at least in part, by stimulating the production of D cyclins. As expected, GGF, which is a potent mitogen, stimulates the production of cyclin D1 and D2 within 4 hours, whereas IGF, which is a poor mitogen on its own, does not stimulate the production of cyclin D1 at all and only weakly stimulates the production of cyclin D2. Moreover, IGF-1 does not increase the stimulation induced by GGF, indicating that this is not how IGF-1 synergizes with GGF for cell-cycle progression.

Ian was now in a position to do the critical experiment. He cultured the cells without serum and in a constant amount of IGF-1, to keep cell growth constant, and added variable concentrations of GGF, to drive the cell cycle at different rates. As expected, cells accumulated much faster in a high concentration of GGF than in low GGF, because the cells go through the cycle faster, even though they are growing at the same rate. This indicates that cell growth cannot be the only thing limiting the rate at which the cells go through the cycle. Because the cells in high GGF are going through the cycle faster but grow at the same rate, they become smaller than cells in low GGF. Cells proliferating in high GGF are smaller than cells proliferating in low GGF. This is the case in all phases of the cell cycle. Cells in mitosis in high GGF, for example, are significantly smaller than cells in mitosis in low GGF. The same is true if one uses flow cytometry to analyze cells in S phase. Thus, cells do not seem to care what size they are when they divide. Their size at division seems to depend on how fast they are growing and how fast they are going through the cell cycle, which depends on extracellular signals, which can stimulate cell growth, cell-cycle progression, or both.

One of the reasons for thinking that there must be a cell-size checkpoint is that proliferating cells in culture tend to maintain a constant distribution of sizes. But experiments in yeast indicate that big cells grow faster than small cells. In a classic experiment done by Paul Nurse and Kim Naysmith in 1972, mutant fission yeast blocked in S phase continued to grow and grew much faster as they became bigger. If that were true for our cells, there would have to be some kind of mechanism, such as a cell-size checkpoint, to ensure that the distribution of sizes remains constant; otherwise, the difference between the smaller and larger cells would get greater and greater, which is not what happens. So do our cells behave like yeast cells, with big cells growing faster than small cells? The answer is no. The increase in volume of cells growing in aphidicolin is linear. The cells add the same amount of volume, and presumably cell mass, per day, independent of their size, which is quite different from yeast cells. Although the cell growth rate is independent of size, it is not fixed. It depends on the concentration of extracellular signals stimulating the growth. If one increases the concentration of serum, growth remains linear, but the slope increases, indicating that the cells grow faster with increasing serum. It seems that, at these concentrations of serum, what is limiting cell growth is not something inside the cell, such as ribosomes. It is the level of extracellular growth signals that limits growth, which may be why yeast cells and our cells are different.

An alternative explanation is that the difference between yeast cells and our cells is an artefact of the way the experiments were done. Ian measured cell volume, for example, whereas Nurse and Naysmith measured protein per cell. Ian therefore repeated his experiments analyzing protein per cell, and he obtained the same results: Schwann cells grow linearly, and they grow faster in 10% serum than in 3% serum. Is it because Ian uses aphidicolin to arrest the cell cycle in S phase, whereas Nurse and Naysmith used a mutation that blocked their cells in S phase? Experiments done by Hudson and Mortimer in the 80's suggest that this is not the explanation for the differences between yeast cells and our cells. They subjected mice to starvation, which rapidly causes the liver to shrink. All the shrinkage was due to hepatocyte atrophy – the cells shrinking in size. When they re-fed the mice, the liver regrew very quickly, due to cell growth, and the cell growth was linear. It seems, therefore, that most, if not all, of our cells grow linearly.

It is remarkable that our cells grow linearly, independent of cell size, as one would expect many things to change when cells get bigger. Big cells have a much reduced surface-area-to-volume ratio compared to small cells, for example, which might cause transport across the plasma membrane to change as cells grow.

Cells grow when macromolecular synthesis is greater than macromolecular degradation. Does linear growth imply that big cells and little cells make macromolecules at the same rate, independent of their size? This seems not to be the case. Cells that have been blocked in aphidicolin for three days in serum are much larger than cells blocked in aphidicolin for only one day, and the amount of protein they make, as measured by ^{35}S-methionine incorporation during a two-hour pulse, is greater. As the big cells grow at the same rate as the smaller cells, they must be degrading proteins faster, which is the case. Thus, the remarkable conclusion is that our cells keep this difference between protein synthesis and degradation constant over a large variation in cell size. It is not known how they do this, but it seems to depend on extracellular signals.

Sympathetic neurons are one of the few animal cell types where we know what controls their size. Dale Pervis and his colleagues showed in the 1970's that the size of these neurons in an adult rodent depends directly on the amount of nerve growth factor (NGF) that they get from the target cells they innervate. If the level of NGF is experimentally increased in the adult, the neurons get bigger, and if the level is reduced (by treatment with anti-NGF antibody), the neurons get smaller. Franklin and Johnson showed that, if one cultures newborn rat sympathetic neurons in NGF and inhibits protein synthesis with cycloheximide, the cells rapidly shut off protein degradation and thereby maintain their size. Thus, protein synthesis and degradation are coupled in some way that is not understood. If the cells are treated with cycloheximide in the absence of NGF, however, protein degradation fails to shut down, and the cell shrinks to a very small size. Thus, the coupling between protein synthesis and degradation is dependent on extracellular signals, in this case NGF, but how NGF acts to regulate this coupling is unknown.

Cell-number control

I now turn to the other main process involved in size control, which is cell-number control. I will shift from Schwann cells, which make myelin in the peripheral nervous system, to oligodendrocytes, which make myelin in the central nervous system.

In mammals, size largely reflects cell number, rather than cell size. We are bigger than mice mainly because we have more cells than mice. We are about 3,000 times the mass of a mouse, and we have about 3,000 times more cells. We study the control of oligodendrocyte cell numbers in the rodent optic nerve because the optic nerve is one of the simplest bits of the central nervous system. It is simple because it does not contain nerve cells, although it contains the axons of the retinal ganglion cells passing back to the brain. It also contains two major classes of glial cells – astrocytes and oligodendrocytes (Fig. 2). The oligodendrocytes wrap around the axons to myelinate them. The question we have addressed is how the final number of oligodendrocytes is determined. Oligodendrocytes, like nerve cells, generally do not divide after they differentiate. They are produced from oligodendrocyte precursor cells (OPCs) that migrate into the optic nerve from the brain early in development. The OPCs divide a limited number of times before they stop dividing and terminally differentiate into oligodendrocytes. Thus, the final number of oligodendrocytes depends on (1) how may OPCs migrate into the developing nerve (we do not know this number, but I suspect that it is small), (2) how many times the OPCs divide before they stop dividing and differentiate, and (3) how much death occurs in this lineage.

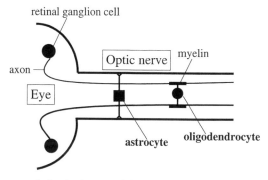

Figure 2. Astrocytes and oligodendrocytes of the optic nerve.

Normal oligodendrocyte cell death during development plays a major part in adjusting the numbers of oligodendrocytes to the number of axons requiring myelination. Ben Barres, who now has his own lab in Stanford, showed that oligodendrocytes are over-produced and that about half of them commit suicide by undergoing apoptosis. We proposed that newly formed oligodendrocytes have to contact an axon in order to survive and that about half the cells find an axon and live and the other half fail to do so and die. This would automatically match the numbers of oligodendrocytes to the number and length of axons. This proposal has been tested in a number of ways. If one cuts the developing optic nerve just behind the eye, almost all of the axons in the nerve degenerate and, as a result, all of the oligodendrocytes die, indicating that the oligodendrocytes need something from the axons to support their survival. If one increases the number of axons but does not change the number of oligodendrocytes produced, then fewer oligodendrocytes die and their number automatically adjusts upwards. Bill Richardson and colleagues showed that if one genetically engineers a mouse so that it produces too many oligodendrocytes, all of the extra oligodendrocytes die, and their number automatically adjusts to normal. Charles French-Constant and colleagues showed that if one engineers a mouse to produce too few oligodendrocytes, the competition for survival signals from the axons is very much less, and so fewer die, adjusting their number upward. Thus, cell death plays a crucial part in controlling oligodendrocyte number.

The final question I want to address is why OPCs stop dividing and differentiate when they do. Most cells in our body are produced from precursor cells that divide a limited number of times before they stop dividing and terminally differentiate. It is true for most blood cells, muscle cells, nerve cells, skin cells, gut cells, and so on. There is not a single case where we know why the precursor cells stop dividing and differentiate. The stopping mechanism is important because it determines how many differentiated cells are produced, and it controls the timing of differentiation. If one studies the development of oligodendrocytes in the rat optic nerve, one finds that the first OPCs migrate into the nerve a few days before birth (gestation is 21 days in the rat), and the first OPCs stop dividing and differentiate in small numbers on the day of birth. If one takes cells out of the embryonic nerve and puts them in culture, the OPCs develop on a similar schedule. The OPCs proliferate, and the first ones stop dividing and differentiate around the equivalent of the day of birth. Thus, one can reconstitute the normal timing of withdrawal from the cell cycle and differentiation in a culture dish. Fen-Biao Gao, when he was a postdoctoral fellow (he now has his own laboratory in the Gladstone Institute at UCSF), showed that purified OPCs from embryonic day 18 (E18) optic nerve behave similarly to OPCs in mixed optic nerve cell cultures: if cultured in platelet-derived growth factor (PDGF), which is the major mitogen for these cells, and thyroid hormone, the first cells become oligodendrocytes on the equivalent of the day of birth. Since these are purified OPCs, this finding suggests that the timing of cell-cycle withdrawal and differentiation is either built into the population of OPCs or built into each individual OPC. The evidence that it may be built into each individual OPC comes from experiments that Sally Temple did when a Ph.D. student (she is now in her own laboratory in Albany, New York). She took a single OPC and put it alone on an astrocyte monolayer, which makes PDGF and survival signals. The cell proliferates to produce a clone. She found that the OPC divides a variable number of times (up to eight times), and then all of the progeny cells stop dividing and differentiate more or less together. To show that this behaviour is built into each individual OPC, she put the astrocytes on the side of a microwell and a single OPC on the base. After the cell divided, she put the two daughters on separate astrocyte monolayers. She showed that, if one cell divided for four days (roughly four times), then so too did its sibling; if it divided for eight days (roughly eight times), so its sibling did too. That established beyond any doubt that built into each OPC is some mechanism that limits its proliferation. The reason that cells divide a variable number of times is apparently that the cells differ in maturity, and immature cells divide more times than do mature cells. Are the cells simply counting divisions and when they get to eight they stop? The answer is probably no. Fen-Biao Gao showed that dropping the temperature to 33°C slows the cell cycle, as expected, but, unexpectedly, it causes the cells to differentiate sooner, after fewer divisions. It seems that the cells are measuring time in some way that is independent of cell division. We therefore call the intracellular program an intrinsic timer.

What is the nature of this timer? We know that it is complicated. Some intracellular proteins that inhibit cell divsion increase over time as the cells proliferate, thereby helping the cells to withdraw from the cell cycle. Other proteins that promote cell proliferation and inhibit differentiation decrease over time, thereby helping the cell to stop dividing and differentiate. One example of the first type of protein is the cyclin-dependent protein kinase (Cdk) inhibitor p27. As shown by Béa Durand when she was a postdoc (she is now at the Pasteur Institute), the mean level of p27 increases as purified OPCs proliferate. For the timer to work, both thyroid hormone and PDGF are required. But even in the absence of thyroid hormone, where most of the cells just keep proliferating and do not differentiate, Béa found that p27 rises and plateaus around the time the cells would have stopped dividing and differentiated had thyroid hormone been present. Jim Apperly showed when he was a Ph.D. student that the over-expression of p27 in purified OPCs causes the cells to come out of division and differentiate sooner, as expected if p27 normally has a role in the timing. Béa collaborated with Jim Roberts in Seattle, whose laboratory was one of the three that knocked p27 out in a mouse. She cultured p27 -/- OPCs at clonal density with PDGF and thyroid hormone so that their timers could work and studied how many divisions the OPCs went through before they differentiated. She found that some of the cells went through an extra division or two before they differentiated, suggesting that p27 is one component of a multicomponent timer.

The p27 knock-out mouse is 35% larger than normal because it has more cells in all of its organs. It has more cells because there is more cell proliferation and not because there is less cell death. This suggests that p27-dependent timers help to stop cell division in many cell lineages. There are p27 homologues in worms and flies, and when these are inactivated, cells go through an extra division or two in multiple cell lineages. It seems that Cdk-inhibitor-dependent timing mechanisms operate in all animals.

How is the level of p27 controlled in proliferating OPCs so that it increases to a plateau as the cells mature? Yasu Tokumoto, when he was a postdoctoral fellow (he is now back in Tokyo), showed that p27 mRNA levels do not increase as the protein increases, suggesting that the level of p27 protein is controlled post-transcriptionally.

An example of an intracellular protein that decreases over time and thereby helps stop cell division is the Id4 protein. There are four known "inhibitors of differentiation (Id)" proteins in mammals. All four are expressed at the mRNA level in OPCs, as shown by Toru Kondo when he was a postdoc (he is now back in Japan). He found, however, that only Id4 mRNA falls over time, while the other Id mRNAs remain constant. The level of Id4 protein falls in parallel with the Id4 mRNA, suggesting that it is controlled transcriptionally. The Id proteins are known to promote proliferation and inhibit differentiation in many cell lineages, and so it is likely that the fall in Id4 plays a part in timing differentiation in OPCs. When Toru over-expressed Id4 in OPCs, differentiation induced by either taking away PDGF or adding thyroid hormone was inhibited, and most of the cells kept dividing. Over-expression of a different Id protein, Id1, had no effect on differentiation. Toru collaborated with Fred Sablitzky in Nottingham, whose laboratory has knocked out the Id4 gene in mice. They studied the development of oligodendrocytes in cultures of Id4 -/- neural stem cells and found that oligodendrocytes developed prematurely. Taken together, these findings suggest that Id4 normally plays a part in timing oligodendrocyte development.

Let me finish by returning to the original question of how it is that we grow to be larger than mice. It is clear that our cells divide more times, on average, than do mouse cells. They divide more times because they divide for longer than do mouse cells, not because they divide faster. The question is why. Is it because intracellular timers are programmed differently in our cells and in mouse cells? Or is it because the extracellular signals that stimulate cell proliferation persist for longer? Or is it some combination of these and other factors? Solving this puzzle is a major challenge for the future. An even bigger challenge is to determine how local controls on growth operate to ensure that most of us end up looking very different from mice.

The Biology of Complex Organisms –
Creation and Protection of Integrity
Ed. by K. Eichmann
© 2003 Birkhäuser Verlag/Switzerland

How the immune system protects the body from infection

Charles A. Janeway, Jr.

Sec Immunbiolology HHMI, Yale University School of Medicine, 310 Cedar St, LH 416, New Haven, CT 06510, USA

Thank you to the organizers inviting me to the occasion of the 40th anniversary of the Max-Planck-Institut für Immunbiologie. And I should like to particularly thank Otto Westphal for spending years purifying my favorite molecule, which is LPS.

I would like to commemorate a former director of this institute, whom I met last at one of those Keystone Meetings. I happened to be skiing with him and he looked pretty content. He died shortly thereafter. This is Georges Köhler, and I don't have to introduce him to anyone, I hope. He invented, together with Cesar Milstein, the hybridoma technique for generating monoclonal antibodies which has really dominated the world of immunology for the last 30 years.

I start with a couple of historical findings. The first is of Karl Landsteiner. He was famous for making antibody responses to any chemical substance he could lay his hands on, and it was rapidly accepted that the immune system could make antibodies to any size or shape. Therefore, it seemed like the immune system was independent of bacteria, viruses or any other infectious agent. But I think that is a fallacy which I call the Landsteinerian fallacy, because any shape can induce antibody production by the immune system specifically during infectious events. And that I think it is wrong because it gives the impression that the immune system is independent of evolutionary pressures from infectious agents.

This is a picture of Eli Mechnikov, a Russian biologist who worked in the late 19th century. He was famous for sticking a thorn into a small hemocyte in a sea star under a microscope. He observed the reaction that was set up by this thorn and I call this the immunologist's dirty little secret. Antibody responses to non-infectious antigens required a mixing of antigen in an adjuvant, which is any substance that enhances adaptive immune responses. The critical ingredient in adjuvants are dead bacteria. So it was based on those two conclusions that I proposed that there had to be more to immunology than just an adaptive immune response. It caused me to think very seriously about innate immunity.

So this is my summary of at least four types of cells that make up the innate and the adaptive immune systems (Fig. 1). I shall tell you how these two systems differ in a minute.

Four types of cell contribute to immune responses: The MHC is the major histocompatibility complex, as I am sure everybody knows. When T cells develop within the thymus, they only see self-peptides plus self-MHC molecules on the

The immune system is made up of at least 4 types of cell

1. Antigen-presenting cells that are activated by innate immune signals.

2. T cells that are referential to self-peptide:self-MHC complexes and control adaptive immunity.

3. B cells that are involved in an idiotypic network as originally proposed by Niels Jerne, and participate in adaptive immune responses.

4. Regulatory or suppressor T cells that act to prevent autoimmunity.

Figure 1. The immune system is made up of at least four types of cells.

thymic cortical epithelium. Self-peptides binding to MHC molecules are critically important to stabilize the MHC molecules. B cells are involved in an idiotypic network as originally supposed by Niels Jerne. Based on data we and others have generated, the receptors on B cells are required for survival. Such B cells also participate in adaptive immune responses. And finally regulatory/suppressor T cells that actively prevent autoimmune diseases.

Innate versus adaptive immune receptors: How do the two systems of innate and adaptive immunity co-exist? Well, first I have to say that the innate immune system is as old as the division between plants and animals. But there are real differences between the operating principles of these two systems (Fig. 2). The receptors for the innate immune system are fixed in the genome. You can trace them from mouse to men and also, as I shall show you in a minute, they are present in plants, invertebrates, and vertebrates. Those molecules are fixed in the genome, and therefore rearrangement is not necessary.

The receptors of the adaptive immune system are encoded in gene segments, and the rearrangement of these gene segments is absolutely necessary for their expression on the cell surface. The distribution of receptors in the innate immune system is non-clonal; they are identical in any class of cells, such as dendritic cells or macrophages. By contrast, the distribution of the receptors in the adaptive immune system is clonal. Therefore, they are all distinct within a class of lymphocytes. The recognition by the innate immune system is of conserved molecular patterns. Sometimes these are called pathogen-associated molecular patterns (PAMPs), such as LPS (lipopolysaccharide). Usually they are products of very complex biosynthetic pathways, and they are absolutely necessary to confer pathogenicity on the pathogen. But products such as a LPS are also found on non-pathogenic enteric bacteria. In this case, we believe that enteric bacterial LPS can be recognized by innate immune receptors, as shown by the septic shock that follows cecal ligation and puncture, which allows bacterial proliferation in peritoneum.

What the adaptive immune system sees is details of molecular structure such as proteins, peptides, carbohydrates and haptens, as originally and elegantly proven

Innate and Adaptive Immunity

Property	Innate immune system	Adaptive immune system
Receptors	Fixed in genome Rearrangement not necessary	Encoded in gene segments Rearrangement necessary
Distribution	Non-clonal All cells of a class identical	Clonal All cells of a class distinct
Recognition	Conserved molecular patterns (LPS, LTA, mannans, glycans)	Details of molecular structure (proteins, peptides, carbohydrates)
Self: Non-self discrimination	Perfect: selected over evolutionary time	Imperfect: selected in individual somatic cells
Action time	Immediate activation of effectors	Delayed activation of effectors
Response	Co-stimulatory molecules Cytokines (IL-1β, IL-6) Chemokines (CXCL8)	Clonal expansion or anergy Growth cytokines (IL-2) Effector cytokines: (IL-4, IFNγ)

Figure 2. Innate and adaptive immunity: how receptors in these two systems differ.

by Landsteiner. In self-non-self discrimination, this is not really important. We think that self-non-self discrimination is largely if not entirely mediated by the innate immune system, and therefore it is perfect or nearly perfect. I have never seen an autoimmune disease that was caused by an innate immune response. And that is probably because the recognition of foreign microbes has been selected to get along with self through the course of evolutionary time. In contrast, we all know that the adaptive immune system is imperfect in discriminating self from non-self because it is selected in individual somatic cells and therefore mistakes happen quite frequently.

The action time of an innate immune response is rapid activation of effectors, whereas in the adaptive immune system one needs about five days until it starts producing antibody and functional T cells. And the response of the innate immune system, at least to LPS, is the expression of co-stimulatory molecules, cytokines, and chemokines. In the adaptive immune system, as we all know, clonal expansion depends on the co-stimulatory molecules, leading first to the production of IL-2 and then switching over to the mature cytokines IL-4 or interferon-γ.

Innate immunity functions via an ancient family of host defense responses. They all basically work on the same principle. They break down the IκB molecule through an intermediate called TRAF-6, and then they proceed through serine/threonine innate immunity kinases (SIIK). These are found across plants, insects and vertebrates, and you can see that there is a lot of similarity between these branches in the evolutionary tree. I am sure, there are plenty of others. But all of these innate responses act in the same way, which is they initiate transcription of host defense responses by activating NFκB by inactivation of the NFκB inhibitor IκB.

Innate Immune Recognition by Toll-Like Receptors

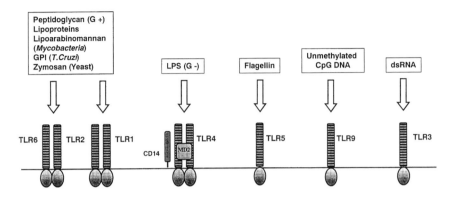

Figure 3. A brief catalogue of known Toll-Like Receptors and their ligands.

Here is a collection of various TOLLs, or TLRs as they are now called (Fig. 3). TLR stands for TOLL-like receptor. We now know that there are ten TLR genes. I got a set of critiques of my book where it was said there are 11. But I do not know of more than 10. TLR2 sees peptidoylglycans, TLR4 sees LPS, TLR9 sees CpG DNA, which is present in eukaryotes in a methylated form, and in prokaryotes in a de-methylated form. TLR3 recognizes dsRNA, which is found in the life cycle of most viral infections, but not in eukaryotic organisms. All of the microbial ligands can activate NFκB, which leads to cytokine production and, in vertebrates, to the expression of the co-stimulatory molecule B7. In insect cells, what happens is induction of a bunch of antibacterial molecules, and these also play a role in host defense. In plants, the genes induced are also involved in protection of the host.

Antigen and co-stimulatory molecules must be present on the same cell in order to activate naive CD4 T cells. We submitted this paper to Nature, which turned it down because the effects were only quantitative. Quantitative by about a thousandfold difference, but still quantitative. We should not have carried out the titration to determine the thousandfold difference. To be serious, one needs to have the antigen presented by the same cell as the co-stimulator; that is, signal 1, which is the foreign peptide:self-MHC molecules recognized by the T cell receptor and its co-receptor CD4, and the so-called co-stimulatory signal. If this is received from the same cell you get T cell activation. If you only have signal 1 and not signal 2, you get anergy instead.

So, what about the adaptive immune system? Well, you have to understand that the lymphocyte as an immune cell was not discovered until Jim Gowans began draining lymphocytes from the thoracic duct, which returns these cells to the blood, and he found that depleting lymphocytes removed all the immunological memory responses from an animal. You could restore these responses by re-infusing the same lymphocytes. So now we can modify the postulates of the clonal selection hypothesis by saying that each lymphocyte has a single type of receptor of a unique specificity, although that is certainly an exaggeration. Interaction between an antigen and a lymphocyte receptor with high affinity for that antigen leads to lymphocyte activation. I should also say that the antigen-presenting cell needs to carry the co-stimulatory molecule. The differentiated effector cells derived from an activated lymphocyte will bear receptors with identical specificities to those of the parental cell from which the effector lymphocyte was derived. That is true except in B cells, which have somatic hypermutation and also class switching. And lymphocytes bearing receptors recognizing self molecules are deleted in an early stage in lymphocyte development and are therefore absent from the repertoire of mature lymphocytes, or controlled by suppressor or regulatory T cells. One can still observe autoimmune diseases which result from an adaptive immune response directed at self antigens by autoreactive lymphocytes.

Having said that lymphocytes directed at self antigens are all deleted in the central lymphoid organs, the bone marrow in the case of B cells and the thymus in the case of T cells, what I want to show you now is, first, evidence that CD4 T cells are self-referential; that is they have to recognize a self-peptide bound to a self MHC ligand, that resembles the ligand on which they were initially selected in the thymus. The MHC molecule never exists by itself; if one tries to make empty MHC molecules, they fall apart right away. We know that the famous MHC restriction phenomenon initially observed by Zinkernagel and Doherty involves a self-peptide. It is self because the thymus is an immunologically privileged site that does not drain materials from the outside. Thus, developing T cells cannot see any foreign peptide in the thymus. They only see self-peptide:self-MHC complexes so that the mature T cell repertoire has to be selected on self MHC:self-peptide complexes. This is true of both MHC class I or MHC class II molecules. CD4 T cells are selected on MHC class II molecules, and CD4 T cells survive in the periphery on complexes of self-peptides and self-MHC class II molecules. And finally, I want to try and convince you, without showing any data, that during activation in the periphery, self-peptide:self-MHC class II complexes contribute to positive signals. These signals are not silent but are always tickling the T cell to basically stay awake and alert. In fact, the well-known autologous mixed lymphocyte reaction actually demonstrates this quite nicely. We are currently trying to find a way to observe this directly.

The next thing I want to tell you is a bit of molecular biology. We used a two-step PCR-strategy on genomic DNA derived from cells from a T cell receptor β chain-transgenic mouse. The α chains were then identified by PCR. You have to know that the α chain is encoded in approximately 50 Vα gene segments and approximately 50 Jα gene segments, so one has to restrict one's attention to the Vα and Jα gene segments used by the T cell receptor from which one derives the transgenic β chain. The only piece of information we derived from this is the length of the TCRα junctions, where we obtained a very interesting pattern.

Normally, we see variants which differ by one amino acid, from 1-6 additions added by the enzyme TdT. However, if one uses mice that are defective in the H-2DM, you have basically a clonal population of antigen presenting cells that lack the ability to exchange the MHC class II invariant chain peptide (CLIP) in the MHC class II binding groove. Nevertheless, these mice allow T cell maturation, and one can see that under these circumstances, one obtains TCRα junctions selected on the CLIP peptide which are nearly all the same length, and there are repeated amino acid sequences, although these protein segments differ in their nucleotide sequence, indicating that they arose independently (Fig. 4).

Next, I would like to discuss the work of a French post-doctoral fellow, Christophe Viret, who analyzed positive selection of 1H-3.1 specificity on H2Mα -/- background. Fig. 5 shows cells bearing the 1H-3.1 TCR transgenic on a normal background, and this is the TCR transgenic on a H2Mα-deficient background. You can see that one gets mainly CD4 T cells in the periphery, and all of these express the transgenic Vβ6 chain.

This is what we see when we examine these CD4 T cells in an adoptive transfer assay. Initially, we measured the survival of Vβ6$^+$ cells in irradiated mice, and one can see that in MHC class II molecule knock-out mice, by 7 or 8 days the CD4$^+$, V$_\beta$6$^+$ are down to less than 10% of the original number, whereas, in an MHC class I knock-out, which carries the MHC class II molecule, I-A^b, on which these cells were positively selected, they actually expand. In normal B6 mice, they also expand. In another experiment we used BALB/c recipients carrying the non-selecting allele, I-A^d, which does not support survival of the 1H3.1 T cells. Thus, we can say quite clearly, at least in these circumstances, one can get survival of CD4 T cells on self-peptide:self MHC class II ligands if you have the right ligand, whereas if you have the wrong ligand, you do not. And many other authors have published experiments like this.

We then took normal cells that had come through an H2M-α -/- thymus. We transferred them into B6 mice, which, as you know, express a whole range of self peptides. The T cells have been raised on the H2M-α -/- mice. When such T cells are transferred to normal B6 mice, they are going to proliferate like crazy against the peptides that are present in normal B6 (I-A^b) mice. But they do not proliferate at all in I-A^b knock-out mice or in H2M-knock-out mice. There is little bit of proliferation on the H2M-knock-out mice, pretty much what we see normally in mice that are raised on a particular set of peptides.

Then we decided that we try the same trick with the B cell receptor in J$_H$ knock-out mice, that cannot make any endogenous rearrangements of their heavy chains. We used two heavy chain transgenes that are related but not identical, and then sequenced the light chains in the immature B cells, we saw five sequences out of several hundreds were V$_K$ 24,25, sequence 80. We then looked at mature B cells, where we saw a four-fold increase in this one V$_K$ 24,25 sequence 80. Using the second, related IgH transgene, we found that a different sequence was selected in the mature B6 cells, V$_K$1 sequence 8 (Fig. 6).

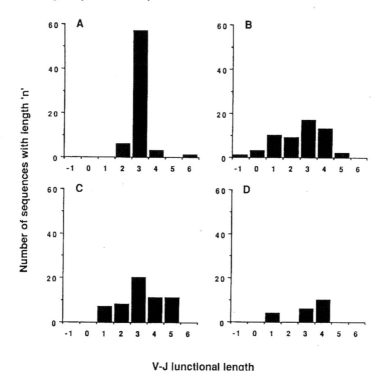

Figure 4. The effects of altering the intrathymic self-peptide repertoire on the mature TCR repertoire. The frequency of different amino acid lengths of the Vα2-Jα4 CDR3α segments of CD4$^+$, Va2$^+$, TCR^{high} thymocytes from H-2Ma$^{-/-}$ mice with the D10 TCRβ chain transgene (n=67) (A); I-A^k mice with the D10 TCRβ chain transgene (n=55) (B); I-A^b, H-2Ma$^{-/-}$ mice with the D10 TCRβ chain transgene (n=57) (C); and H-2Ma$^{-/-}$ mice without a β-chain transgene sorted for CD4$^+$, Vα2$^+$, Vβ8.2$^+$ (n=20) (D). V-J junctional lengths are listed by potential N-encoded codon additions. With this system, the D10 CDR3α segment would be +1.

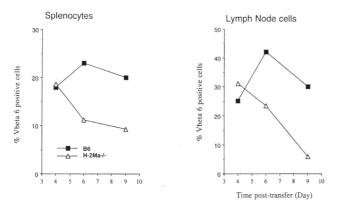

Figure 5. The survival of 1H3.1 T cells in C57BL/6J and C57BL/6J with a deletion of H-2Mα -/- that present a limited number of peptides mainly restricted to the MHC class II invariant chain peptide CLIP.

B Cell Receptor Positive Selection

Heavy Chain	Light Chain Immature	Light Chain Mature
Vh 186.2	Vκ24/25 s80: 5	Vκ24/25 s80: 19
Vh J558	Vκ s82: 6	Vκ1 s82: 22

Figure 6. Summary of data on light chain selection in mature versus immature B cells.

What that says is that the idiotypic network, if it works on anything, obviously works on the primary repertoire and is not, as Mel Cohn has argued passionately, and very correctly I believe, as a regulatory mechanism. Instead, we believe that the idiotypic network of Niels Jerne is a developmental mechanism that controls the structure of the primary B cell repertoire.

Finally, I want to show you evidence that T cell receptor recognition of self-peptide:self-MHC molecule recognition is not only important for positive intrathymic selection and for onward survival of T cells, but also plays a role in T cell activation. The argument is complex, so I want to take you through it one step at a time.

It begins with what I have long thought about T cell activation, namely, that conformational changes upon binding to activating complexes of foreign peptides bound to self MHC molecules, as shown by several individuals, including Rolf Zinkernagel and Peter Doherty, activate T cells. In an attempt to characterize the T cell receptor, we raised many monoclonal antibodies, using the Köhler-Milstein technique, against a single T cell clone. By comparing binding to the receptor and T cell activation, we were able to divide these antibodies into two types, which we called high potency and low potency. Both sets of antibodies varied in binding affinity over a broad range, but when this was factored out, they fit neatly into these two types. The high potency mAbs could cap the T cell receptor, while the low potency antibodies could not. The high potency antibodies could co-cap the CD4 co-receptor, whereas the low potency antibodies could not cap the TCR. To cap the T cell receptor with the low potency antibodies, we had to use an anti-Ig antibody, yet the CD4 molecule remained free on the cell surface. Thus, although I had evidence for a conformational change in the TCR, I could not validate it.

I needed to understand what was going on, and the answer appears to be provided by the co-author of my immunology textbook, Paul Travers, who used measurements made in an optical device with great sensitivity called a BIAcor. This machine used a chip with one protein mounted on it, which is bound by a second protein in a solution that is passed over the chip.

These sorts of experiments, which I am just going to tell you about in theory, are based on studies of Alam, Gascoigne and Travers. They are BIAcor experiments, and they were first done at room temperature, 25˚C. They compared stimulating and non-stimulating peptides, and they did not observe a great deal of difference

(Fig 7). You can see the stimulating peptide had a half-time of 30 sec. while the non-stimulating peptide that bound the TCR as shown by its ability to drive positive selection in fetal thymic organ culture, has a slightly shorter half-life of 20 sec. Then they realized that the biological results they were studying were performed at 37°C, so they re-ran the same reagents at 30°C. In this experiment, they put the MHC molecule peptide complex on a chip and flow over T cell receptors, the result they get is quite remarkable. They get a doubling in the amount of T cell receptor binding, and the half-time is about twenty-fold increased. Whereas, if they used a related peptide that could drive positive selection, what they observed is basically a half-time that is increased by 1 sec., which is within experimental error. At least, in this case, there is evidence that there is T cell receptor dimerization.

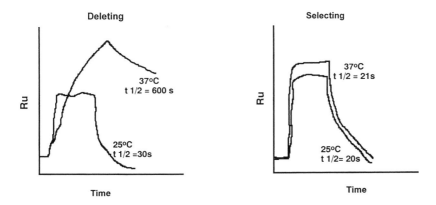

Figure 7. Schematic showing the dramatic effect of temperature of soluble TCR specific for a deleting peptide and a selecting peptide. Science: Alam, Travers and Gascoigne, Immunity (1999).

This next figure is a schematic drawn to illustrate what I believe is going on in this experiment (Fig. 8). There are two peptides illustrated, one self-peptide and the other activating peptide. The activating peptide induces a conformation that binds to normal TCR confirmation. When one has an agonist-induced confirmation in the periphery, one gets activation, as you know. In the thymus, one gets negative selection, as everybody knows as well. But the thing that intrigued me was this: in the periphery, one observes sustenance if the T cells recognize self-peptide:self:MHC complexes, whereas in the thymus, one observes positive selection on self-peptide:self-MHC complexes.

And this also tells us the same thing. It is designed to illustrate the case of the 1H3.1 TCR. This TCR is the only one we have tested, but it gives very anomalous results. It gives positive selection when you have pure activating peptide in MHC class II binding groove. This is the only example that we have that sending a signal to a T cell requires conformational change in the T cell receptor. To produce a pure agonist peptide in the MHC class II binding groove, we have to use a very complex maneuver. What we actually observe is very strong positive selection. I thought about this for months and this is the only way that I can put it together (Fig. 9).

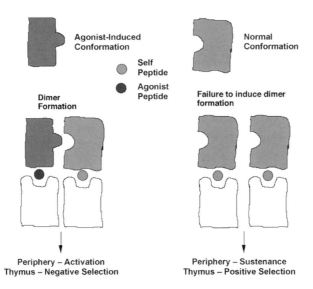

Figure 8. Schematic showing the dramatic effect of temperature of soluble TCR specific for a deleting peptide and a selecting peptide. Science: Alam, Travers and Gascoigne, Immunity (1999).

If T lymphocyte development requires self-ligands and persistence also requires self-ligands, why are cells that are receiving activating signals from self so rare? One would think that self-ligands would also trigger these T cells to respond in an autoimmune fashion. And the answer seems to be a forgotten proposal of Richard K. Gershon, which was that suppressor T cells can counteract the actions of effector T cells that mediate all autoimmunity.

I have a post-doctoral fellow, named Margaret Bynoe, who approached this question in a very interesting way. We knew that epicutaneous immunization does not allow Th1 responses. Organ-specific autoimmune diseases are mostly Th1-mediated diseases, both in mice and in people. So what she did was to apply patches to the skin, after soaking them in Ac1-11, which is the peptide that is recognized by our MPP-TCR transgene. She has carried out her experiments in transgenic mice, which may be a little unfair. She found that a dose of 10 μg of Ac1-11 would very strongly suppress this response and prevent active immunization. If instead, she soaked her patches in PBS, she got a very strong response to immunization with Ac1-11. This procedure involves two steps, patching the mice two times, and then removing the patch and immunizing the same site with Ac1-11 and adjuvant. The data show that she gets a dose range and the optimal dose is 10 μg/patch. She also demonstrated that the suppression of EAE is specific.

Now the next question was: Could CD4[+] CD25[-] cells transfer protection to naive recipients. The story here is that suppressor cells in many antigen-non-specific systems are entirely CD4[+] CD25[+]. She transferred CD4[+] CD25[+] or CD4[+] CD25[-]

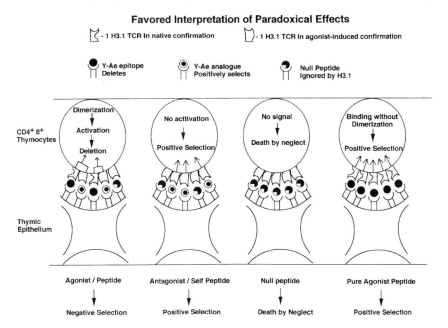

Figure 9. Schematic of the effects of the results of having no self-peptide ligands which leads to positive selection of 1H3.1 T cell receptor-bearing cells.

cells, at varying doses. She saw the strongest suppression using 10^6 CD4$^+$ CD25$^-$ cells. As far as CD4$^+$ CD25$^+$ cells are concerned, she also observed some suppression, but it was no where near the suppression she observed with CD4$^+$ CD25$^-$ cells.

She also invented a weaker kind of spontaneous EAE, as the spontaneous EAE seen in mice that can not rearrange their receptors is so fast that you cannot intervene in it. Using the same transgene together with a Cα -/-, she observed spontaneous disease, in which the disease takes longer, but eventually one does get disease. If you use Cα knock-out of the MBP TCR-transgenic mice, patch them with 10 µg soluble Ac1-11, you can completely suppress this spontaneous disease. So my vision of this experiment is that we hope to be able to eventually prepare a strip of skin patches to apply to the forearm of a child; it may take at least a hundred years to turn this prevention of organ-specific autoimmune disease into reality. By delivering antigens across this epithelium, we should be able, in principle, to protect against diseases like diabetes, multiple sclerosis, and many other autoimmune diseases that arise during a lifetime.

Thank you Otto, Klaus, Michael, Thomas, Rolf and Davor for your kind invitation, and to the audience for your attention.

The Biology of Complex Organisms –
Creation and Protection of Integrity
Ed. by K. Eichmann
© 2003 Birkhäuser Verlag/Switzerland

Biological curiosities: what can we learn from them?

Jacques Miller

The Walter and Eliza Hall Institute, PO Royal Melbourne Hospital, Melbourne, Victoria 3050, Australia

Many of you would agree with me, especially after this morning session, that experiments in biology are not always straightforward. If you want to understand bodily functions you have to perform experimental manipulations that may not always mimic the normal physiological process. What I intend to do this afternoon is to discuss how valuable to biology are the conclusions that we can draw from such experiments. And I shall illustrate this with examples from my own work – examples that are scattered throughout the last 40 years.

Many immunologists will know that prior to 1961 the thymus was not considered to have any immune function, primarily for two reasons. The first was that adult thymectomy did not interfere with immune responsiveness. This fact led a group of American workers to conclude that "the thymus gland does not participate in the control of the immune response". The second reason was that thymus lymphocytes unlike lymphocytes from other lymphoid tissues were not able to adoptively transfer immunity to appropriate recipients. This fact led Sir Peter Medawar to conclude that "we shall come to regard the presence of lymphocytes in the thymus as an evolutionary accident of no great significance". But from observations that were made on mice in which the thymus had been removed immediately after birth, I was able to conclude that the thymus does have a crucial immune function. This might have been expected of an organ which is rather large in the infant, and which atrophies after puberty and involutes to thin shreds of fibrous tissue containing lymphocytic islands in old age. However, it seemed odd to me that thymus functions should cease abruptly after puberty. One way to check this was to ask whether the adult thymus was still required, in order to allow the immune system to recover after it had been damaged by irradiation. Although the experimental design I used to determine this is very simple, many people, even some immunologists, still have difficulty in understanding it. As I will use this experimental design throughout this talk, let me explain it now. Irradiation destroys the hemopoietic and lymphoid tissues. The animal dies unless it is rescued by giving bone marrow, because bone marrow has stem cells that can differentiate to hemopoietic cells and to lymphoid cells. So the question I asked was, if we take out the thymus of an adult mouse and then subject the mouse to heavy doses of irradiation and inject bone marrow, will this mouse recover its lymphoid system properly? The technique I used was therefore adult thymectomy followed by heavy doses of irradiation and an intravenous injection of bone marrow. There was no recovery of immune functions in adult thymectomized irradiated and bone marrow protected mice. The control was a mouse that had a sham operation, and then irradiated and given bone marrow. I concluded that because the thymus was present in the control, recovery of its immune system did take place. Yet, at that time, Burnet expressed reservations about the significance of these observations. But before I go on to cite Burnet's objection, let me just remind you of one very important experiment that proved the existence of two major subsets of lymphocytes, T and B cells. By using adult thymectomized, irradiated and bone marrow-protected mice as recipients of semi-allogeneic cells from different sources, my Ph.D. student Graham Mitchell and I proved that there were two types of lymphocytes. We did this at a time before any markers were available to distinguish these two cells and we therefore used, as markers, the antigens determined by histocompatibility differences.

Burnet expressed reservations about the significance of results obtained in such "biological monstrosities as inbred mice that were thymectomized, lethally irradiated, and protected by injection of bone marrow from another mouse". How could such complicated experimental manipulations ever relate to normal physiology? For never in the history of human ailment had any disease been seen to mimic the effects of adult thymectomy, irradiation, and bone marrow protection. In the 29 years that elapsed since this statement was made, I have seen only one other example of such a vituperative term being used to denounce a scientific entity. That was in the case of the wavicle or wave particle of Quantum physics. "The idea that something can be both a wave and a particle defies imagination and there is nothing in every day world that remotely resembles such a monstrosity". So, my thymectomized, irradiated and bone marrow-protected mice stand in very good company since Quantum physics which has the wavicle as one of its central tenet, Quantum physics which has caused a collapse of common sense, has nevertheless been extremely successful in its everyday practical application, for example in microelectronic devices such as diodes in our television sets.

But let me come back to Burnet's biological monstrosities. What have we learnt from their use, if anything!

(1) As stated above, I used it to show that the thymus continues to exert its function in adult life. Is that correct? This seems to be true, as recent work not utilizing such a monstrosity, but sophisticated molecular biological techniques has shown that the aging thymus still rearranges T cell receptor genes, and thus still has an effect in adult life. Investigators have analyzed T cell receptor excision circles in human thymus. These are generated as the T cell receptor genes rearrange so that part of the DNA gets excised and circularized. They conclude that their presence in aged people shows that their thymus is still active. Thymic T cell generation was detected even after 60 years of age. So it looks as if the older thymus does continue to exert its original immune function.

(2) The use of thymectomized, irradiated and marrow-protected mice enabled Graham Mitchell and I to prove the existence of the two major subsets of lymphocytes which, of course, is well established, now that we have so many markers to distinguish between these two sets of cells, markers that we did not have at the time we performed these experiments.

(3) Thirdly, the use of these mice allowed Zinkernagel and colleagues to show that MHC restriction is imposed intrathymically on developing lymphocytes. They did this by using adult thymectomized irradiated and bone marrow-protected mice with the additional bonus of putting in a thymus graft. And these experiments have been repeated many times, even using organ cultures of thymus *in vitro*.

(4) Fourthly, the use of adult thymectomized and irradiated rats was instrumental in showing that autoimmune disease can arise in lymphopenic animals. There must therefore exist immunoregulatory T cells, as Janeway mentioned this morning. It will be very interesting in the future to determine whether those immunoregulatory T cells come out of the thymus after 3 days, as mice thymectomized 3 days after birth do get autoimmune diseases. That is something that will have to be investigated thoroughly as soon as possible.

It seems therefore that we have learnt quite a lot from the use of what Burnet referred to as a "biological monstrosity".

I am certain that Burnet would consider a mouse producing in its pancreas an allogeneic MHC molecule or a chicken protein as another type of biological monstrosity or biological curiosity as I would prefer to call it. When transgenic animals became available, we created such a transgenic mouse. The idea behind this work was to ask the question, what would happen to T cells that come out of the thymus and that have specificity for an antigen that is tissue-specific and sequestered. I will briefly summarize what we have done with these transgenic mice and will ask whether what we have found is relevant to biology.

The RIP-K^b B10.BR mouse (haplotype kk) was first used. It expressed the allogeneic MHC class I molecule H2-K^b, or K^b for short, in the β cell of the pancreas. We predicted that this mouse would get diabetes. Why? Because K^b-specific cells that were generated in the thymus would, when they come out, encounter the K^b antigen on the surface of the β cell, and hence would become activated and kill those β cells. The mice did indeed develop diabetes. However, as we showed in detail in several publications, against all expectations the diabetes that ensued was not associated with any lymphocytic infiltration in the pancreas. The pancreatic islets were shrunken and there were very few insulin-producing cells. They were disorganized and the glucagon and somatostatin-producing cells were scattered throughout instead of being positioned peripherally as in normal islets. But, the most important finding was that the absence of lymphocytic infiltration. So, as immunologists, should we have disregarded our results as a laboratory artifact not worthy of being further studied? Fortunately, we decided to investigate and I will now summarize our further findings. We found that these mice were tolerant of K^b-bearing skin. Tolerance was presumably due to the fact that very few molecules of the K^b transgene were present in the thymus – in fact they were detected only by the reverse transcriptase polymerase chain reaction. We next took thymuses from RIP-K^b mice, that as we showed expressed only a few molecules of K^b, and transplanted them to thymectomized, irradiated, nontransgenic bm1 mice protected with bm1 bone marrow. bm1 mice express-$2K^{bm1}$ instead of H-$2K^b$. Tolerance to K^b was induced in these thymus-grafted recipients. Such few molecules of K^b were therefore sufficient to induce tolerance in nontransgenic cells differentiating in the transplanted thymus. Furthermore, with the help of Bernd Arnold and his group in Heidelberg, we could show that only T cells with high affinity for the transgenic antigen were deleted intrathymically. T cells that had low affinity for the antigen escaped thymic censorship (escaped negative selection). Yet these cells ignored the pancreatic islet transgenic antigen, unless they were provided with exogenous help in the form of IL-2. In another experiment using different transgenic mice, called Eμ-K^b, that expressed small amounts of K^b only in the medullary epithelial cells of the thymus, we showed that tolerance to K^b occurred. So, what did these experiments reveal?

(1) They added weight to the notion that negative selection deletes predominantly high-affinity T cells. Low-affinity T cells that escape negative selection generally pose no problem unless extra help is provided.

(2) Our work using Eμ-K^b transgenic mice was followed up by others, notably by people in Heidelberg, to show that thymic medullary epithelial cells are not only the primary site for negative selection but also synthesize many self-antigens previously thought to be tissue-specific and sequestered. How so few molecules of self-antigens, present in only so few medullary epithelial cells, can induce self-tolerance, remains a mystery that begs to be elucidated.

Let me now pass on to a third biological curiosity. We used a transgenic mouse that expressed the major peptide of the ovalbumin (OVA) molecule, SIINFEKL, as a membrane-bound molecule in the β cells of the pancreas. Now, unlike the H-2K^b mice, the expression of this transgene in β cells had no deleterious effect on β cell function. But like the H-2K^b mice, the thymus also expressed the transgenic antigen. So we thought we could manipulate these mice further, in other words create a double Burnetian biological monstrosity. Let me explain this experiment in some detail. We used a transgenic RIP-OVA mouse that presents SIINFEKL on the β cell of the pancreas, and also in the thymus. We first thymectomized these mice, to remove the thymus that does express the SIINFEKL transgene. We grafted a syngeneic but non-transgenic thymus that does not express the SIINFEKL molecule. We then heavily irradiated these thymus-grafted mice and, as we have done in the past, protected them with bone marrow coming from another set of transgenic mice (OT-1 mice), which have CD8 T cells that expressed a high affinity class I-restricted T cell receptor directed to the SIINFEKL antigen. This receptor utilizes the Vα2 and Vβ5 segments of the T cell receptor genes and the T cells harbouring these can therefore be identified. After the mice had recovered for several weeks, we removed the thymus graft and the lymph nodes draining the pancreas to determine what had happened to the T cells with high-affinity receptors directed to SIINFEKL. In the thymus graft, there was no deletion of those CD8 T cells which expressed the transgenic T cell receptor specific for SIINFEKL. This makes sense, because the thymus graft did not express the transgene, as it came from a non-transgenic mouse. In the lymph nodes draining the pancreas, however, there was a loss of the specific T cells: 20.3% in the control nontransgenic mouse versus 3.3% in a transgenic mouse that expressed the SIINFEKL in the β cells of the pancreas. So there was a loss of the high-affinity T cells in the lymph nodes draining the site of transgene expression. Why was there such a loss? Further experiments showed the following:

(1) The transgenic self-antigen in the pancreas migrated via antigen-presenting cells from the pancreas to the draining lymph nodes.

(2) There it was recognized by OT-1 T cells coming from the thymus graft.

(3) The antigen-presenting cells were bone marrow derived, had a short life span and cross presentation of antigen by these was essential for CD8 T cell activation.

(4) As a result of activation the OT-1 cells proliferated.

(5) After some proliferation these activated OT-1 cells underwent activation-induced cell death, a type of feedback mechanism that ensures that activated T cells eventually do not continue to proliferate.

(6) Diabetes occurred only when high numbers of OT-1 cells where given, numbers that are much higher than the precursor frequency found in a normal mouse.

(7) Diabetes also occurred only when very large doses of islet antigen were cross-presented. Low doses did not induce diabetes unless the antigen-expressing cells had been destroyed or damaged.

So what does this all mean?

Well, first of all I would like to say that we have no way of telling whether the phenomenon that I have just described occurs in a normal, unmanipulated, nontransgenic mouse. But the very fact that this mechanism exists, suggests that it may be a way of dealing with potentially autoimmune T cells that have escaped thymus censorship. We might for example suggest the following. Self-reactive T cells, that have escaped thymus censorship, circulate and enter regional lymph nodes. There, they are immediately confronted by the self-antigen carried by antigen-presenting cells. They therefore become activated as soon as they enter the lymph nodes. As a result some proliferation takes place but activation-induced cell death occurs before the T cells can induce damage. By contrast, T cells specific for a foreign antigen coming out of the thymus are not immediately confronted by the foreign antigen, and will therefore build up their numbers to a normal frequency of T cells specific for that foreign antigen. So, it seems that we have a mechanism that removes self-reactive T cells in the periphery, although we cannot prove that it does exist in a normal mouse. One very illustrious scientist whose name I will not reveal, and who was unable to reproduce this cross-presentation mechanism in his model, presumably believes that such a mechanism is totally unphysiological. But even if it is a laboratory artifact, I do not feel we should shelve it, because it might tell us something about the biochemical pathways that occur during cross-presentation of exogenous antigen to CD8 T cells. It might tell us something about the signaling events that emanate from dendritic cells, whether they be immunogenic or tolerogenic. It might perhaps tell us something about peripheral tolerance or it might tell us something completely unexpected and novel.

As a general conclusion I would say that no matter what immunological monstrosities or biological curiosities we have engendered by our experiments, we should on the one hand not jump to conclusions by stating that the results we have obtained is what is happening in a normal animal. On the other hand, we should think twice about discarding our line of work, in case further examination leads us to discover something that may be unexpected. And I think Shakespeare said this already 500 years ago when he stated "There is a tide in the affairs of men which taken at the flood leads on to fortune – omitted all the voyage of their life is bound in shallows and in miseries".

The Biology of Complex Organisms –
Creation and Protection of Integrity
Ed. by K. Eichmann
© 2003 Birkhäuser Verlag/Switzerland

Quality control of immune self non-self discrimination

Philippe Kourilsky

Institut Pasteur, Immunology Department, 25, rue du Docteur Roux, F-75724 Paris cedex 15, France

Ladies and Gentlemen,

I am extremely pleased to be here, I thank you for the invitation. The friendly atmosphere which I feel shows that the spirit of this place is part of its history.

Today, I shall review a number of immunological questions some of which have been addressed brilliantly by Charly Janeway and Jacques Miller. Since I shall deal with these questions under a special angle and mostly for the purpose of discussion, I have decided to take off all my slides. Relevant facts have been given by my colleagues. Others are, of course, in the literature. Some come from my own laboratory in Paris. I will start by giving an outline of an approach which we developed to analyze the repertoires of T cells. I will point out one part of this work in which we measured the dimension of the naive T cell repertoires. How many different T cells do we have in the body? To answer this basic question, we took sets of primers specific of the various T cell Receptor (TCR) gene segments and analyzed the size of the PCR fragments. Then, we sequenced to exhaustion whatever was in a given size class-exhaustion also refers to the post-doc doing the work. The number of distinct sequences increases as more sequences are gathered but tends to reach a plateau. I will not go in detail through this methodology. The key point is that in the mouse, we found that there are about 2 million different TCRs, borne by some 200 millions T cells. In humans, the figure we obtained is about 30 million. Thus, although the body weight of a human is about 5000 mice, the complexity is not 5000 fold higher. More recently, we studied a knock-out mutant mouse lacking terminal desoxy-nucleotidyl-transferase (TdT), which generates junctional diversity. We ended up with a figure of about 200 000, so the increase in diversity provided by this enzyme is about ten fold.

The reasons why I report this work is that it induced me to check the numbers in order to see whether the diversity of T cells is capable of covering the diversity of the outside world. T cells recognize peptides displayed by MHC molecules. Calculations can be made in several ways. They are relatively easy because peptide epitopes are 9–10 amino acids long. There are 20 possibilities of each position but anchor residues must be removed. D. Mason has done such estimations. We have done it too. Let me tell you the result, which reveals a major discrepancy. In mice, the number of T cells is too low to match the number of possible foreign epitopes. This goes worse if one goes from man to mouse, because there is a 15-fold difference, and even more so in the TdT knock-out mouse. The latter, even though it has not been released and studied in the wild, is apparently healthy. We do not know if it can resist in all situations but basically it is a good mouse that suffers no real problem. In so-called more primitive animals, frogs and others, the diversity may not be very high either. The point is that there is a discrepancy between the diversity of T cells and the complexity of the outside world. Whenever the basic immunologist tries to immunize, he usually succeeds. Attempts to find holes in T cell repertoires have almost always failed. Thus, the adaptive immune system appears to be complete or nearly complete. Therefore, this implies that the recognition of peptide-MHC complexes by the TCR is not so accurate and actually quite degenerate. Now I shall make a conceptual statement which will really start my talk.

We know that the immune system occasionally fails. Some of these failures translate into autoimmune disorders. The point which I want to make is that there are two ways of looking at this. First, one can say that, because the immune system is basically cross-reactive, it makes mistakes because of cross-reactions. It is, therefore, an imperfect system. Many mistakes must be related to molecular mimicry, which fools the system and makes it react against self. This is one way of looking at it. But there is a different conceptual approach. One can also say that because the system is fundamentally cross-reactive, it has actually been selected to cope with cross-reactions and, therefore, the problem might not reside in the failure of the system itself. To cope with these cross-reactions, the immune system must have evolved some kind of quality control devices. Then the argument is turned around: The idea is that what happens in some of these autoimmune disorders is that the quality control devices have failed. This is the hypothesis. I would like you for 30 min. or so, to wear quality control glasses, and try to think along this concept, which I summarize as follows: there should be at lot of quality control in the immune system to secure the discrimination between self and non-self. So let us take the intellectual attitude in which we are looking for possible mistakes, and for the possible mechanisms which correct them. Then we can see whether it makes sense to look at autoimmune disorders as being at least partly related to failures in quality control.

I would like to emphasize that, although this approach has been used by immunologists, it has not, in my opinion, been explored extensively. In contrast in other fields of biology where it is being used very productively. Take DNA synthesis and maintenance, which involves a large number of quality control devices. Take protein synthesis assembly, folding and intracellular transit. One way of looking at a number of neurological diseases caused by protein aggregation within cells is to think of them as failures of the quality control devices which normally get rid of these aggregated products. I wish to stress that the rate of errors in biological systems can be extremely high, a feature which some of us immunologists tend to forget. I would like to remind you of the fact that 30-40% of nascent proteins attached to ribosomes are inadequate and are channeled to the proteasome to be degraded. Let me add something which has been alluded to by Lewis Wolpert and Rudolf Jaenisch in a different context. In my opinion, we also tend to forget that there are many stochastic variations in biological processes. And they have to be dealt with. There is good evidence for this when one looks at variations in gene expression in individual, though identical bacteria. If one deals with multicellular organisms, one has to realize that there must be a very large number of what I would call internal polymorphisms. These include the random mutations which take place in our cells within our own body. There are other types of phenotypic polymorphisms related to variations in the expression levels. More generally, if one piles up the stochastic variations within a complex and interactive system, one may end up with very broad variations in the output. We have to keep this in mind.

Regarding the immune system, there is another type of polymorphism which is what I would qualify as external polymorphism. This indeed originates from HLA, or the MHC, the major histocompatibility complex, which varies enormously from individual to individual. As it shapes the immune system, it implies, of course, that the immune system varies enormously from one person to another. But the point is, precisely, that the performance of the immune system of different individuals is remarkably similar. In my view, this implies that the immune system is built in such a way that it is permissive to errors, polymorphic variations, that is internal and external polymorphisms. This, in turn, means that when thinking about self non-self discrimination, we should not think exclusively to the elements which constitute the somatic self as it is seen by the immune system. But, perhaps, we should focus more on the rules which govern the discrimination. The argument is that these rules are common between the individuals and guarantee a regular outcome, irrespective of these enormous individual variations.

Let us go back to the notion of the question of Quality Control, or QC. Of course, a number of quality control mechanisms have already been identified within the immune system. Some of them are very obvious. Giving credit to Charly Janeway for his discoveries on the role of innate immunity, we can surmise that innate immunity is also a quality control device of adaptive immunity. It is obvious that any device which involves two signals bears in itself a quality control system. One may also argue that the mere existence of two recognition systems, one by B cells and one by T cells, reflects a quality control system. I find it is very surprising that during evolution, somewhere in the fish times, both MHC, T cells and B cells apparently emerged all at the same time. Of course, we do not have all the records that we would need to be absolutely sure that this is exactly what happened. Nevertheless, we may speculate that it could not happen otherwise, because a B cell system by itself would be too dangerous since it would make too many errors. Scanning the protein antigens into fragments recognized by T cells may be a way to restrict antigen mimicry at the antibody level. We know that many infectious agents do handle antigen mimicry in a very efficient way. The existence of NK cells is also a typical internal QC device, since NK cells recognize cells which have lost the expression of MHC class I.

QC is also visible at the molecular level. For instance, what we know today about the so-called immunological synapse reveals a number of checkpoints. When certain checkpoints are not made, the core synapse dislocates. I noted that in your talk, Charly, you claimed that the innate immunity was "perfect" – you said close to perfect! I would argue, on the contrary, that any system makes mistakes. Even a specific receptor makes mistakes, and innate immunity does so as well. It can also be mistaken by molecular mimicry, regarding certain sugars, or combination of sugars. Innate immunity itself might be subjected to a number of quality controls which have yet to be found, as well as the interface between innate and adaptive immunity. Think of the non-classical MHC molecules which detect the formylated peptides found at the NH_2 terminus of bacterial proteins. They exist in mitochondrial proteins as well, so that there is a potential for confusion. The most important quality control system which has been found to date is probably the regulatory T cells. They are now extremely popular. It is remarkable – as Jacques Miller has mentioned – that the critical data have been there for about 30 years. The 3-days thymectomy experiments, in which mice systematically develop autoimmune disease, have been described in 1969. It is only in the last few years, despite the important work of Sakaguchi, in particular, that this has become accepted by the community of immunologists.

One aspect which deserves being emphasized is that, thanks to regulatory T cells, tolerance can be dominant and not only recessive. In addition, I wish to quote the striking experiments performed by several groups, including Waldmann's and Coutinho's. They have claimed tolerance mediated by regulatory T cells to be infectious in the sense that one cell is able confer to another cell its regulatory phenotype. This is probably an extremely important aspect. Several observations also indicate that the acquisition of tolerance should be seen as a developmental process, with a window at about day 3 in the mouse, during which the first wave of self-educated T lymphocytes and future regulatory T cells exits the thymus and then provides some kind of memory of initial events during life. This is still somewhat speculative but this is potentially very important.

On top of this, there are so many similarities between B and T cells that I wonder whether regulatory B cells are going to be found. In my opinion, anti-idiotypic B cells might well be functional homologues of regulatory T cells.

Let us move on to the question of self non-self discrimination. We shall focus on the rules rather than on the elements. Focusing on the elements does generate a number of difficulties. Together with Jean Michel Claverie, in 1986, we formulated the so called peptidic self model. We had realized at the time that any cell had to be covered with thousands of peptides associated with its class I MHC molecule. In this way, a cell displays on the outside a sample of what it has within. Thereby the notion of immune surveillance had to be expanded according to the existence of allo-reactivity and anti tumor specific cytolytic T cells (CTL) could be easily explained.

Along this view, the immune self includes the catalogue of self peptides. We are in the era of genomes, we have catalogues of genes, we have algorithms which allow to predict, for an individual who bears a certain set of MHC molecules, the set of self peptides which should be displayed. Thus, one should be able to build this catalogue of somatic self peptides. However, this is not sufficient because what is pertinent for the immune system is what the immune system reacts with. Therefore, what is needed is not this catalogue, it is the subset of this catalogue which is immunologically pertinent. This raises the issue that we do bear in ourselves a number of elements which are foreign to the immune system. This is why we can hope to vaccinate against some cancer cells. There is another problem with this type of molecular definition of the self. The pertinent catalogue of self peptides must be seen as composed of a stable core surrounded by a cloud of variants. This is because the peptide presentation system has stochastic features. Take the rarest peptides. Obviously, some of them must be presented as two or three copies on the cell and another time there is no copy or one, and this can vary with time. Therefore, we have another difficulty when strictly dealing with the elements per se, namely that they can hardly be defined as a constant set. So let us turn to the rules.

What are the rules? Well, they include the rules which govern the formation of the repertoires of naive cells, as well as the rules which govern the triggering and development of the immune response from this naive repertoire. The point I want to make is that these rules have something in common. They all involve the setting of avidity thresholds. This is true both of the selection of T cells in the thymus and of the activation of T cells during the triggering and the development of an immune response. It now appears likely that, as predicted by Grossman and colleagues, in particular, T cells modulate their thresholds during their life. This is now well established for T cells which mature in the thymus and migrate out of it. The question of whether they modulate their thresholds individually during their life as a function of their experience is linked to that of memory T cells. Accordingly a T cell clone is not a clone anymore, in the sense that clones of T cells do have the same TCR but are functionally heterogeneous, because the thresholds are adjusted at different levels in different cells. If this is correct, the affinity maturation of B cells with somatic mutations may well be paralleled by an avidity maturation of a T cell as they experience antigen. Regulating these avidity thresholds must be, and this is where I wanted to come to a major target for quality control.

Let us now follow this route taking as a readout the occurrence of autoimmune disorders. So, what do we know about autoimmune disorders? There are two fields of observation and experimentation. One relates to the spontaneous emergence of autoimmune disorders in humans and certain mice. The other deals with what researchers have produced in terms of mutant mice in the laboratory, which, sometimes unexpectedly, show autoimmune symptoms.

It is well known that the general phenomenology and ontology of human auto-immune disorders includes the genetic background, mainly HLA, but also many other genes. It is also related to the environment, but nobody knows exactly what environment means, probably infections, perhaps intestinal flora. For example, in certain animal models of diabetes, the disease develops only if the animal facility is reasonably dirty. What does it mean in molecular and cellular terms remains to be elucidated. Of course, at this point comes the issue of molecular mimicry, produced by infectious agents, either as a trigger or as an amplifier of disease. In particular, a derivative of a drug used in humans deregulates the threshold of positive selection in the mouse thymus, creating a lupus-like syndrom. This is, therefore, one example in which deregulating the selection thresholds correlates with an autoimmune syndrome. Another question deals with the inflammatory context which can develop within an organ. What I want to emphasize is that, when one reviews the literature, one finds a large number of mutant mice, some of which generated for irrelevant purposes, which develop autoimmune disorders. Quite interestingly, a zoo of knock-out mice with autoimmune syndromes has been produced. Thus, the inactivation of certain glycosylating enzymes can cause immunological disorders. Some of these mutations may affect the immunological synapses. The driving concept currently is that glycosylation as such does not impact on the specificity, but modifies the kinetic parameters which regulate immune reactions. The idea is that sugars restrict the movement of molecules, and modifies certain kinetic parameters, which translates into kinetic features. Apparently, almost each time that people have knocked out a molecule involved in either T cell activation, B cell activation, B and T cell co-operation, they observed some autoimmune disorder. There is still another beautiful experiment, performed by Wakeland and his colleagues, in which they analyzed the lupus-associated loci in lupus-prone mice and found that the CD2 molecules were playing a significant role.

The importance of this approach, in my view, relates to its possible extrapolation to man. In humans, it is likely that null mutations will be rare. What is happening in the mouse is that, rather unwillingly, because researchers are not always interested in immune disease, people are generating a repertoire of candidate genes which might show up in humans not as null mutants but as polymorphic variants however. The number of these genes is potentially quite large. Therefore, a lot of distinct combinations of these polymorphisms may generate the same phenotype that ends up in autoimmune disease. Going backwards, this implies that the same phenotype can be obtained by multiple combinations within a large subset of mutations. If that happens to be the case in humans, the family studies which try to identify the genes in diabetes and other diseases might produce different results, which will all be correct, in that the identified genes belong to a broad pool out of which various combinations contribute to disease.

I will conclude by saying the notion of quality control is interesting because it stimulates a certain way of thinking which, I believe, should be more present in the brains of immunologists. It can contribute to an understanding of immune specificity which is quite different of what some of us were taught in their youth, namely, that immune recognition is very specific, like a key-lock system. This is absolutely not true. The specificity of the immune system is built in a completely different way, by incremental adjustment of multiple checkpoints. This is the case of cognitive systems: It is fair to say that Burnett's theory which in immunology has been so productive and useful but has now reached its limits because it left absolutely no space for physiological autoreactivity. Because of this, it cannot accommodate dominant tolerance, it does leave space either for quality control thinking. To conclude, I shall say that, as we make use of a more cognitive view of the immune system, we get closer to the time when we shall learn from complex networks in general. There are people working on complex networks in mathematics and physics, in academic but in banks as well. The internet is a complex network. Very important notions may come out of this. For example, in certain types of networks, the reliability of the complex network is inversely correlated to its sensitivity to attack. Think in these terms of the internet and its attack by viruses. Think of it in terms of the immune system, and mistakes in the immune system, could there be a trade-off between autoimmunity, autoimmune disorders, and sensitivity to infections? The stakes are very high – Michael Sela knows better than anybody here – the stakes in terms of therapy are enormous, I am confident that in the near future we will witness spectacular progress in these areas.

Part II: Ceremony

The Biology of Complex Organisms –
Creation and Protection of Integrity
Ed. by K. Eichmann
© 2003 Birkhäuser Verlag/Switzerland

About the history of the MPI for Immunobiology

Otto Westphal

Chemin de Ballalaz 18, CH-1820 Montreux, Switzerland

Ladies and Gentlemen,

When Michael Sela said how old or how young we are, I realize that one of our great friends in London is now 101 years old, Walter Morgan, who would have come if he would not be in a bad shape at the moment. He used to say that we are extinct vulcanos and remain so for long, but anytime ready to break out. Michael Sela is a good example.

Dear colleagues and friends, let us go back for 65 years to the middle of the 30ies. At that time I had just finished my studies in Science and Chemistry and was looking for a thesis. I visited Professor Karl Freudenberg (1886 - 1983) in Heidelberg who at that time was the chief of the chemical institute and one of the most well-known stereochemists of his time. When I asked him whether I could work with him, he started to speak about Karl Landsteiner. Charles Janeway already praised Landsteiner; I must do it again because he is really one of our great scientific grandfathers. Freudenberg had just studied a publication in the Journal of Experimental Medicine about the three stereoisomeric tartaric acids which Landsteiner had coupled to protein, showing that each of the epitopes is inducing antibodies which allow to distinguish the three very closely related structures, the D-, the L-, and the meso-tartaric acid. Freudenberg continued he had an idea that the human blood groups A, B, and O may be something like a D-, L-, and a meso-configuration. "Would you like to work on that?". Completely ignorant as I was I said yes, yes. From this moment on my whole scientific life was preprogramed, I had to learn carbohydrate chemistry on the one hand, and to learn serology, immunology, on the other.

I was very lucky to be in Heidelberg; there was a pupil of Paul Ehrlich, Professor Hans Sachs, 1877-1945 (Fig.1), the director of an institute for experimental cancer research with a special department of serodiagnostics.

From him I learned about the basis of what I needed for my work. He also had an excellent library, maybe one of the best in Germany where I tried to keep up with the latest news in the field of immunology. Here I learned about Oswald Avery's work on specific pneumococcal polysaccharides, their purification and immunology. In 1938 appeared a publication, again from the Rockefeller Institute in New York, of Walther F. Goebel, describing the identification of the serological determinant of the highly infectious type III Pneumococcus as the disaccharide cellobiuronic acid. Goebel coupled this type III determinant to protein applying Landsteiner's

Figure 1. Professor Hans Sachs (1877–1945).

azo method. Injection of the artifical antigen into rabbits led to the production of specific type III antibodies which protected the animals against thousand-fold amounts of highly virulent type III pneumococci – at that time in the USA number one of causes of death by infection.

Goebel's fundamental observations did not find the general attention that they should. The reason was obvious. In Germany, Karl Domagk had just introduced the sulfonamides. In England, penicillin was found as highly active antibiotic around 1938. And it was shown that pneumococcal infections, especially, were best treated by either sulfonamides or by penicillin. Thus, synthetic antigens against pneumococci had only theoretical interest. But I always remained fascinated by Landsteiners chemo-specific antigens and the azomethod of their production – by Oswald Avery's work on the pneumococcal polysaccharides, their purification and serology – finally and consequently, again in the same institute, Walther Goebels artificial type III antigen and its highly anti-infective immunological activity! For my whole scientific life, and in my lectures at the university I tried to "infect" young students by telling them this great story, a classical basis of immunochemistry. They used to call me "Zucker-Otto".

In 1942 I got a call to Göttingen to the University, where they were interested in these subjects, and they asked me to give the chemistry lectures for medical students and at the same time try to introduce interested students into immunochemistry. I accepted and got my own little institute, and the first coworkers for scientific endeavours appeared. One was Botho Kickhöfen and another one was Dr. Otto Lüderitz. Both are here today with us. We met in 1944, and we never separated. All what followed, we practically did together.

In Göttingen in the Institute of Hygiene was a Professor F. Schütz interested in making vaccines against Rickettsia, typhus fever. Rickettsia were an infectious virus very much spread over the eastern front of Germany; many people got infected and it was a great need for a vaccine. Rickettsia could not be cultivated in large amounts easily. However, there was a serological cross-reaction with a Gram-negative bacterium, Proteus OX19, the Weil-Felix reaction, which patients who got through typhus fever developed in their serum. Schütz had an idea, one might be able to use Proteus OX19 as a basis for a vaccine against typhus fever. We started to extract *Proteus*, and at this occasion the hot phenol-water extraction method was introduced. In the following it became the method of choice for extracting such kind of material – found to be lipopolysaccharide – from the surface of bacteria. By mere chance, one of our coworkers put this material in a mortar trying in the usual way to make a very fine powder. A tiny amount of material escaped into the air and was probably inhaled by this gentleman. In the evening he got 42° C of fever and everybody thought it was a very serious thing; but after 2 days he was fine again and felt beautifully. So, what we had discovered there, was that the active material from *Proteus* was the bacterial pyrogen.

In the German clinics artificial fever was still a very important means of curing diseases, non-specific therapy. Therefore we thought we should go ahead with such kind of work. But the war ended, scientific research was very difficult for years.

Around that time I got a letter from the Wander Company in Bern, well-known for their dietetic product Ovaltine. A friend of mine, Dr. William Foerst, Editor-in-Chief of the widely distributed journal "Angewandte Chemie", had recommended me to the Wander's . They relied on the fact that the Allied Forces had declared all German patents to be no more valid. Who ever wanted to apply them, was allowed. In the Wander company somebody told them "Make Ehrlich's Salvarsan; syphilis is still a great problem". But they did not realize perhaps that in the meantime penicillin was on the way. Anyway, I got a request from the Wander company, would I be ready to move to the Swiss border, create a little institute with a small team of researchers to cooperate with them. I was able to convince them that we should work on artificial fever and material to produce it. We would make it ready for clinical use. In 1946 it was decided that we move from Göttingen to Säckingen on the Rhine river, not far from Basel.

We found a building in Säckingen that had been a fabric for woodpreparations. It had been given up by the proprietor and could be hired by the Wander company (Fig. 2). In this building two floors of laboratories, a library and a seminar room were installed (Figs 3 and 4).

Being directly at the Swiss border we were able to invite foreign speakers for lectures. So came Woodward, for instance, who had got the Nobelprize for the synthesis of cortisone. For him it was the first visit to Germany after the war; he gave us a seminar for a few priviledged people. We had a great advantage of being in a situation were the contact with foreign scientists became better and better. Colleagues from Basel, like T. Reichstein, and friends of the CIBA, and also from Zurich, like V. Prelog, became our guests.

With the Wander company we started to work on pyrogens. Soon we felt that we

Figure 2. The improvised Wander-Institute in Säckingen, 1948.

Figure 3. O. Westphal and O. Lüderitz in the Wander Laboratory, 1949.

should have clinical contact. I must say, all my life I tried to find clinicians willing to cooperate with scientists. In the near Freiburg was Professor Ludwig Heilmeyer, the famous internal medical professor, widely known as expert in iron metabolism. I told him what we were doing. And he said "o.k. you come immediately, you can work in my clinic, I give you one of my assistants, he will supervise volunteer experiments on artificial fever, and help to analyze its mechanisms". Now, I should tell you at that time, 1947/48, here in Freiburg, most of the university laboratories had been destroyed at the end of the war (Figs 5 and 6).

Figure 4. The seminar room in the Wander Institute in Säckingen.

Figures 5 and 6: View from the tower of the cathedral on Freiburg before and after November 1944.

Now, improvisation played an enormous role; but it brings people together. The more difficult things are that you want to do, the better you cooperate of course. It was a very stimulating atmosphere in which we started studying injections in volunteers, most of them medical students. The scientific chief of Wander said "if you cannot pay the volunteers, because the German Mark is not interesting, you can offer them, maybe, Ovaltine which we are producing. We can also introduce sugar and butter from Switzerland. And every volunteer can get a nice package". We followed that suggestion, and you cannot imagine how many people, especially before Christmas and before holidays, were showing up and wanted to join the experiments. I can assure you that in the several hundreds of pyrogen experiments, none of the volunteers ever had serious side effects, some reacted with high fever. Our clinical colleague, Dr. Walter Keiderling, supervised the trials in the best way, and Prof. Heilmeyer held his powerful hand over our activities. The Wander company carefully produced and controlled the ampoules with our highly purified *Salmonella* lipopolysaccharide (Pyrexal) selected and standardized for the trials.

The following picture is a documentation of a voluntary experiment (in this case on myself) that we did with Prof. Heilmeyer (Fig. 7).

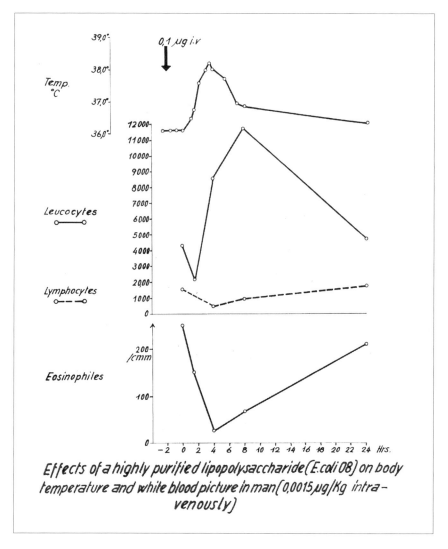

Effects of a highly purified lipopolysaccharide(E.coli 08) on body temperature and white blood picture in man (0,0015 µg/Kg intra-venously)

Figure 7. Volunteer experiment after i.v. injection of 0.1µg LPS.

The i.v. injection of 0.1 µg totally of lipopolysaccharide (LPS) creates fever, leucopenia followed by leucocytosis, and a characteristic eosinopenia. This is what medical people at that time called an alarm reaction, the body's "Aha" on something irritating its equilibrium, which you can induce with totally 0.1 µg of bacterial LPS. Observing such fantastic reactions we were very stimulated of course, to go ahead along these lines and to know more about the underlying mechanisms. I am very pleased that Dr. Philippe Kourilsky from Paris is here today. In the thirties his parents were working about the biology and immunology of endotoxin together with the great pioneer in the field, Prof. André Boivin at the Pasteur Institute (who died from cancer in 1952 much too early).

When we installed our little institute in Säckingen, we needed a lot of material for the building; but nobody wanted to be paid in Marks. Also here the Wander company said "If they like to have sugar or butter, or whatsoever, we will introduce them". So, we paid the whole institute in Säckingen by goods. Once I calculated how much that would have been in money; the whole institute was erected for less than 10.000 Swiss Francs – rather a cheap and a very nice institute! For quite some time people said, you have the best institute in Germany.

Now slowly international contacts came up and people wanted to cooperate. The first big congresses started, one on microbiology in Rome in 1953. Dr. Wander said "If you would like to send Dr. Otto Lüderitz, he should go, later he can tell us what they discussed there". So, Otto Lüderitz went to Rome, and there he met for the first time many experts in our field with whom to cooperate was our greatest desire. From this moment on we were in contact with Anne-Marie Staub of the Pasteur Institute in Paris and with Fritz Kauffmann of the Statens Serum Institute in Copenhagen. Later, as Michael Sela told you, we got very close to colleagues of the Weizmann Institute. I must say, at that early time very productive relations. In this sense we had indeed one of the finest post-war institutes in Germany.

One day, Dr.Wander came and said "Are you not too far away from centers of science, you are here in the country. Should you not go nearer to a university? If you like, I shall build you a nice institute. Would you like that idea?" Think that today somebody would offer you to build an institute for you wherever you would like it! You can easily imagine what our answer was. Dr. Wander decided that a research institute should be built here in Freiburg, exactly on the ground where you are here today (Figs 8 and 9).

In 1956 we started to work in this beautiful institute. We had many guests, one of the first was Fritz Kauffmann, the one who had introduced the Kauffmann-White-System of *Salmonella* species which defines every *Salmonella* strain by its serological speficities expressed in numbers 1, 2, 3, 4, 5, etc. We liked to call that system Kauffmann's *Salmonella* telephonebook. Now, together with Anne-Marie Staub and Kauffmann we coordinated these numbers with clear chemical structures. In years of immunochemical research we worked on that question and we were finely able to link sugar and oligosaccharide structures with specificities 1, 2, 3, ... what they really mean chemically.

At that time we had many guests, and I started a guest book from which one can reconstruct many of the visits. On October 28, 1958, Professor Karl Thomas from

Figure 8. The Wander Institute in Freiburg under construction in 1954.

Figure 9. The Wander Institute completed in 1956.

Göttingen came to visit us. He was a specialist in research on Silikosis at the Max-Planck-Gesellschaft in Göttingen. I showed him around. After some time he asked "How are you organized here?". I answered "Well, I think like a Max-Planck-Institute". Later, Karl Thomas told me he went home, he slept and he woke up and suddenly it came in his mind: why not a Max-Planck-Institute? He went to the Max-Planck authorities and said "Should we in Germany not have this institute as a Max-Planck-Institute?". They started negotiations with Dr. Wander. Of great help was the Professor of Physiology in Bern, Alexander v. Muralt, a friend of the Wander family. The idea was also supported by many friends of the Max-Planck-Gesellschaft and its senate, like Erich v. Holst and Richard Kuhn. It was positively decided. I guess this was on that date which you fixed for the forty years when it was decided to transform the Wander-Institute into a Max-Planck-Institute.

Figure 10. Herbert Fischer (1921–1981).

In our negotiations with Professor A. Butenandt, President of the MPG, we argued that we cannot have a Max-Planck-Institute only with Immunochemists. We need an experienced biologist who comes right from the basis of work on non-specific immunity and its induction by the Aha-reactions of the animal and human body. So came Herbert Fischer from Frankfurt (Fig 10). We had met him in Basel at a congress where he spoke about fibrinolysis. We knew that our lipopolysaccharide is a very strong inducer of fibrinolysis, which makes it of course additionally very interesting for medicine. With Herbert Fischer came Dr. Paul Gerhard Munder (1928–2002) who became a renowned expert for the activation and mode of action of macrophages.

The institute was opened in March 1962. We were, of course, proud how many people joined and were interested. It was a very stimulating meeting, now about fourty years ago (Figs 11-13).

The institute continued to intensify international cooperations and also – important – its relation to the university. Professor Peter has kindly described how much the institute and the university cooperated and really tried to create a new image of immunology in Germany. As I told you, at the beginning, the interest in vaccines lay only in the field of viruses, no more very much in bacteria, because there were good antibiotics and chemotherapy. Generally, interest in immunology and immunological research was not on a high level. Only when in 1955/56 in England and elsewhere modern immunology suddenly came up, it became clear how much we need it. In Germany our group and a few others were for quite some time the only ones, especially due to the fact that in the 30ies before World War II, many of the very good immunologists had to leave the country. Also my dear teacher, Prof. Hans Sachs, had to go at the very last moment in 1938. So we were really very few left and, so far, it was important to convince people that immunology is something very necessary and exciting.

Figure 11. Inauguration of the MPI for Immunobiology, March 1962; from l. to r.: Anne-Marie Staub, Lieselotte ter Haak, Walter Morgan and Karl Freudenberg.

Figure 12. At the inauguration of the MPI, Otto Warburg and Karl Thomas.

Figure 13. At the inauguration of the MPI, from l. to r.: Walter Gerlach, Otto Hahn, Otto Westphal, Anne-Marie Staub and Walter Morgan.

As to the history in relation to the Wander-Institute and the Max-Planck-Institute for Immunobiology in Freiburg I like to mention the following:

1957   F. Hoff's book on fever, immunotherapy and non-specific therapy appeared, describing in detail clinical observations with our pure bacterial lipopolysaccharide.

1958   In August an Endotoxin Conference in Freiburg was organized by the Wander Institute. About 50 experts from foreign countries joined. Since 1958 every year worldwide, a conference is being organized.

1965   The 15th Colloquium of the Society for Physiological Chemistry in Mosbach near Heidelberg devoted the 3 days totally to the subject immunochemistry.

1967   Foundation of the "Gesellschaft für Immunologie";

1968   its first congress took place in Freiburg. O.Westphal was elected president for the first 10 years. The number of members continuously grew over one thousand. Today, I think it is true that every German university in its relevant faculty has at least one chair related to immunology.

1971   Creation of the "European Journal of Immunology", with Dr. Botho Kickhöfen as the first Editor-in-Chief for many years to come. The secretary of the Journal was installed in the MPI in Freiburg where it still works.

1988   Foundation of the "International Endotoxin Society", several of its presidents being earlier members of the MPI, Freiburg.

What we did first in those years of the sixties was to clear up lipopolysaccharides, endotoxins, of *Salmonella* and *E. coli*. Most of the results are published. In the following Fig.14 the general structure of lipopolysaccharides – from smooth and rough strains – is schematically shown. As you can see, the oligo- or polysaccharides are all bound to what we called lipid A. It is this ubiquitous lipid A of LPS which our team found to be responsible for all endotoxic reactions.

Most of the successful coworkers in the field worked with Otto Lüderitz who had the central laboratory for the education of young students to become endotoxinists. Ernst Rietschel and Chris Galanos joined for years. They are with us today. They, and many more elaborated the formula of lipid A (Fig.15). Together with Japanese friends, especially Prof. Tetsuo Shiba in Osaka, we were able to synthesize *E. coli* lipid A and to show that the synthetic material was identical with the natural. Indeed, it exerts all the activities of endotoxin. The result completed an important chapter of endotoxin chemistry, and it opened many more chapters in the biochemical and biological fields. Since these exciting days, one can say that research on endotoxin was again and again increasing, also concerning its medical aspects.

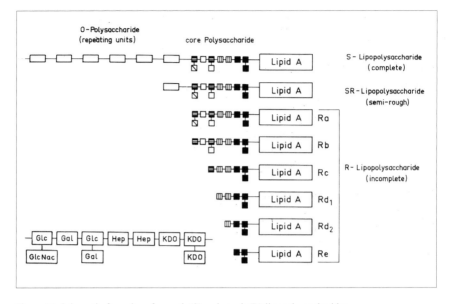

Figure 14. Schematic formulas of smooth (S) and rough (R) lipopolysaccharides.

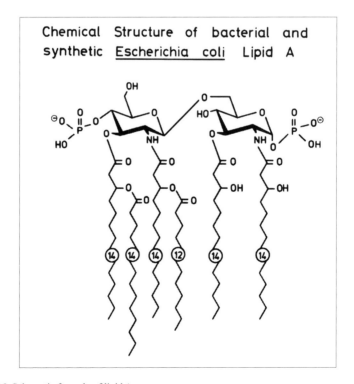

Figure 15. Schematic formula of lipid A.

Figure 16. Dr. Georg Wander (1898–1969).

In closing, please, allow me one remark. I think it is fair to say that we would not be together here today if one man, by virtue of his strong belief in the contribution of basic and applied research to the welfare of mankind, Dr. Georg Wander, had not made his own clear decision to create and finance a Wander Research Institute in 1956 which several years later, with his full support, was transformed into the Max-Planck-Institute for Immunobiology. I think we owe this wonderful man the greatest respect and we should keep his memory very highly (Fig 16)!

Ladies and Gentlemen, it is an old principle of the Kaiser-Wilhelm/Max-Planck-Gesellschaft to build institutes around scientists who are devoted to new and promising fields of research that are still not regularly established at universities. When this Max-Planck-Institute was founded in Freiburg, 40 years ago, immuno-biology and related fields in Germany were in such a state. In the meantime many aspects of specific and non-specific immunology and of immunochemistry, and this is also true for endotoxin research, strongly developed. There are now many investigators who are intensely working on questions around these topics. Taking this into account, the directors of the institute, twenty years ago, became free for a re-orientation of the institute's main lines of research. Klaus Eichmann will certainly elaborate on that. It remains to me and my old friends who gathered here today to congratulate the institute and to wish you the same highly collegial and friendly cooperation that we had in our first institute. It was wonderful, everybody could co-operate with whom he liked. He did not need to ask; nobody knew where the one department ended and another one began. When a doctoral fellow came and said "can I go to this colleague because I need to learn this method?" I said "why had you not already been there"? In this period it was Dr. Wander's spirit who started the institute, and I wish you that this good spirit will keep for the next forty years to come. Thank you very much.

The Biology of Complex Organisms –
Creation and Protection of Integrity
Ed. by K. Eichmann
© 2003 Birkhäuser Verlag/Switzerland

The second twenty years of the MPI IB

Klaus Eichmann

Max-Planck-Institut für Immunbiologie, Stuebeweg 51, D-79108 Freiburg, Germany

It is very hard to talk after Otto Westphal!!
Thank you very much, Otto, for your thoughtful remarks about the origins and the early phase of this institute. We have come a long way since then. I can illustrate this using the consecutive numbering that the institute gives to employees. Everyone gets a number and, of course, Otto Westphal got number one. When I came 20 years later in 1981, which is half of the 40 years that we now celebrate, I was given number 563. When I looked up the most recent employment here, this person received number 2330. This illustrates how many people went through this institute in the first 20 years and in the second 20 years.

Another matter illustrating the dramatic expansion of this institute is space. Otto showed you a picture of the first laboratory building of the Wander institute before 1961. Many buildings have been added since it became a Max-Planck-Institute in 1961 and the newest part of the institute was opened in the beginning of the 90ies, which almost doubled the previously existing laboratory space. By space and personnel, the institute expanded about two and a half times since I came. Of course, it tells us something about science in the old times and science in the new times. Otto told us that he has been together with some of his collaborators since 1945. This rarely happens now.

Nevertheless, some colleagues stayed with the institute almost for the entire length of time. One of them is our photographer Lore Lay. Some of the pictures I will be showing are from the book about 40 years of the Max-Planck-Institute that Lore Lay composed. I regret that we cannot give it away for free, but you can buy it for a very small contribution during the course of the evening. It is a nice book, and I thank Lore Lay very much for all the work that she did and I also thank everyone who sent pictures. This was the reason why I was asking you in my invitation letter to send pictures, and everyone who sent pictures will find him/herself in the book. It should be a nice souvenir. In my short presentation, I will show you some of the pictures from this book and I will comment on some main events and crossroads in the course of the development of this institute since 1981.

On Fig. 1 you see my group as it was during the early years after my arrival. Some of the people had already been members of the institute, like Hans Ulrich Weltzien, who is actually still a member of my department. Others came with me from Heidelberg and stayed on, for instance Markus Simon and Ingrid Falk, but most others joined me and went away like comets, people that are with you for a short time and then they move on. Some of the latter kind have actually acquired

Figure 1. Abteilung Eichmann.

rather prominent positions in German science, like Stefan Kaufmann, who is now
the head of the Max-Planck-Institute for Infection Biology in Berlin, or Jörg
Epplen, who took the chair of human genetics in Bochum, or Frank Emmrich, who
is now professor of Immunology in Leipzig. I cannot mention everyone. It was a
fine group, and it was very nice to begin with this group here.

At this point I do have to mention Herbert Fischer. The other two former directors,
Otto Westphal and Otto Lüderitz, can speak for themselves, because they are here.
Herbert Fischer cannot. He was acting director of this institute when I came in
1981, and unfortunately he died 6 months after my arrival. Very tragic. He had
made my beginning here very pleasant and easy, and I do want to acknowledge and
mention this very warm welcome that he gave me. I am still grateful to him. He was
the discoverer of the Immune Fabrik, whatever that is. He also left some very infor-
mative time-lapse videos about T cell-macrophage interaction which illustrate cell
interaction events that we only today begin to understand. He also was a wonderful
person to discuss science with.

At the time of my arrival the main focus of this institute were gram-negative
bacteria, notably the compound endotoxin as well as other bacterial products, not
exclusively but mostly from a viewpoint of structure. My own interest was in T
lymphocytes and how their specific defense potential was activated and regulated.
While I knew that the work done was internationally recognized, I did not think at
the time that it really deserved to be called immunobiology, and I felt that it was my
task to make immunology the central discipline at the institute. Conversely, I
remember a conversation that I had with Westphal during which he asked me: Do
you really find these small little cells interesting? So there certainly was a gap in
opinion about what is interesting in science. This was the time when immunology –
how I understood it – began to blossom as an independent discipline internationally
and also in Germany, mainly owing to the advent of molecular biology and genetics

Figure 2. Empfang G. Köhler.

as major means of approach. Mouse genetics began to develop, and the institute had just established an animal facility that was among the most advanced in technology at an international scale. It thus was my dream to develop the institute into a center of immunology with a focus on mouse genetics, although I could not foresee at the time the full impact that mouse genetics was to have in the years to come. Retrospectively, this approach was among the most productive in immunology and other biological sciences in the last 20 years, so one can probably lean back and conclude to have taken a good decision.

In this endeavor I was fortunate to be able to attract Georges Köhler to join this institute in 1984. During the same year he obtained the Nobel prize and on Fig. 2 you see the reception that we gave at the occasion of the Nobel prize in 1984. Of all the people I do want to mention Benno Hess, who was the Vice-President of the Max-Planck-Society at that time. I also want to mention two women, one is Ursel Weltzien and the other is Barbara Eichmann. Both of them have been and still are very important women in maintaining a human and social atmosphere at this institute, even with the dramatic expansion. The Nobel prize reception was a rather serious celebration. We had of course also some more relaxed celebrations (Fig. 3). In addition to science, there were good times also on other grounds.

A very important period in the development of this institute began shortly thereafter when the former prime minister of the state of Baden-Württemberg, Lothar Späth, became very excited about the Nobel prize of Georges Köhler. As a result of this excitement, he donated to the Max-Planck-Society a large sum of money earmarked for our institute, which made it possible to build a new laboratory building, the one in which this celebration is now taking place. There was a competition for the design of the building and the architect, Mr. Hecker, won the prize (Fig. 4). Of course we were sitting for hours on end to discuss with the architects how the building should look like.

The opportunities provided by this new building were manyfold and actually

Figure 3. Fröhliches Fest, Weihnachten 1989.

wonderful for this institute. As a result, it was possible – on the one hand – to establish novel scientific programs, and – on the other hand – to continue to optimally accommodate the research activities of the times of the Ottos, and Herbert Fischer, to maintain the established scientific staff and at the same time bring in new people. Retrospectively, maintaining endotoxin research at this institute was a lucky stroke. While at the time it seemed at a stage where merely some loose ends still needed tying, it has now reconquered the interest of immunologists owing to the discovery of TOL receptors, needless to say by mouse genetics, and their pivotal role in innate immunity.

The avenues of research to be newly established were a matter of much discussion at the time. The Max-Planck-Society likes to support research which is not already well established in German universities, and all of us felt that immunology, particularly as it restricts itself in a narrow sense to the handling of foreign antigens, had begun to be quite well represented in German universities in those years. Conversely, mouse genetics had become a general approach and was conquering many disciplines in biology. Consequently, we felt that it would be very good for the institute in its future development to broaden the scope of research activities beyond immunology in a narrow sense. During all the discussions that were held, infectious disease biology and developmental biology crystallized as the two major alternatives. Georges Koehler's dream at the time was to identify and characterize the hemopoietic stem cell, whereas my own ideas became increasingly influenced by the devastating infections of the developing world, including malaria and AIDS. Finally, and to no one's surprise, developmental biology was the field that was chosen. What happened thereafter is a good example that scientific developments may often refuse to follow prospectively planned avenues. In the course of producing series of cytokine knock-out mice that needed testing for their resistance to infection,

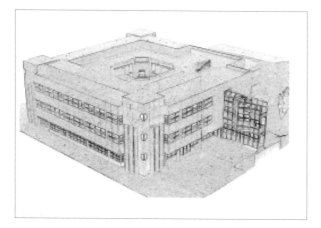

Figure 4. Blueprint of the new laboratory building.

Georges had to establish an array of infectious disease models and became an infectologist of some sort. Conversely, my own interest turned to T cell development in the thymus, mainly due to a chance observation of my student Christiaan Levelt in the early 90ies. Today let me say that we are very happy having made the decision for developmental biology. The richness of mutual information that comes from both sides fertilizes our thinking and makes our science much more enjoyable and interesting. The developmental biologists that came were Davor Solter and Rolf Kemler and established their departments in the beginning of the 90ies (Figs. 5, 6). Rolf Kemler was actually one of those who had been previously at this institute as a student.

This new building enabled us to further expand the institute by establishing junior research groups. They were named the Hans-Spemann Laboratories after the

Figures 5 and 6. Davor Solter (left), Rolf Kemler (right).

Figures 7 and 8. Thomas Boehm (left), Michael Reth (right).

famous developmental biologist who worked in Freiburg during the early 20th century. The present Spemann group leaders are Matthias Hammerschmidt, Ursula Kling-müller, and Viktor Steimle. Spemann group leaders are young promising scientists which are given excellent resources, they can work entirely independently at the institute for 5-6 years, and establish their scientific identity. The present one is already the second generation of Spemann group leaders.

The most recently appointed of the presently four directors is Thomas Boehm (Fig. 7). He came in 1997. He actually succeeded Georges Köhler. Thomas Boehm took over his department, renaming it Developmental Immunology. Georges Köh-ler died in 1995 tragically and untimely. Recently it became possible to establish another junior research group, which we named Georges Köhler Junior Research Group, and Jörg Kirberg is the first leader of this group. He joined the institute a few months ago.

Last not least, another expansion became possible, due to the long-standing and cordial relationships that we hold with the Faculty of Biology of the University of Freiburg. After many months of discussion, it was agreed that the university would establish a chair of molecular immunology and the holder of this chair would work and have his laboratories at this institute. Together, members of this institute and of the university would teach a curriculum in molecular immunology to biology students. Michael Reth is the first holder of the chair of immunology here, also a former member of this institute (Fig. 8). Freiburg is only the second university in Germany where students of biology can choose immunology as a main subject. As you can see, the institute has not only become a much larger body than it used to be, it also developed into an interdisciplinary research center with a unique approach in research and teaching.

Thank you very much for your attention.

The Biology of Complex Organisms –
Creation and Protection of Integrity
Ed. by K. Eichmann
© 2003 Birkhäuser Verlag/Switzerland

Short history of immunology in Freiburg

Hartmut H. Peter

President, German Society for Immunology,
Klinikum der Universität Freiburg, Abt. Rheumatologie u. klin. Immunologie, Hugstetter Str. 55,
D-79106 Freiburg, Germany

Dear guests, dear colleagues, dear friends, dear Prof. Westphal, dear Klaus Eichmann,

it is a great pleasure to address a few words of gratitude and friendship to this audience. On behalf of the members and the board of the German Society for Immunology I first want to congratulate the "Max-Planck-Institut für Immunbiologie" to its 40th anniversary. Over the last 40 years there has always been a special relationship between our German Society for Immunology and this Max-Planck-Institute. To be more precise: the founding father of this institute, Otto Westphal, also founded our Society, and therefore he may be considered the great godfather of German immunology. He actually founded this society in 1968, and for a long time, at least for 10 years, about 50% of our members came somehow from Freiburg. Actually in the early years the German Society for Immunology was considered sort of a country club of the Max-Planck-Institute für Immunbiologie. It took about 20 years until full national and global expansion of our society took place. This happened in the mid 80ties, when the German Society was charged to organize the World Congress of Immunology 1989 in Berlin. Then actually immunology as a science spread over our country; many other institutes were founded and contributed to the scientific development of our discipline. Today our Society is a flourishing enterprise with more than 1500 members (www.immunologie.de). We are proud of our roots, which lay in this institute and also in this city.

You have heard all about the history of the Max Planck Institute für Immunbiologie by Otto Westphal and Klaus Eichmann. Their story tells you also the history of immunology in the second part of the 20th century in Freiburg. But some of you who are not from here, might ask "Was there any immunology in this city before?" Was there something like this spirit of immunology that has been alluded to by Philippe Kourilsky? Be sure, there was and I want to tell you during the next 15 minutes a few highlights of this remarkable history of immunology in Freiburg during the 19th and early 20th century. You will learn that immunology in Freiburg goes back exactly 200 years including the 40 years of the Max Planck Institut für Immunbiologie.

The first most remarkable immunologist in Freiburg was Johann Matthias Alexander Ecker (Fig. 1). He was surgeon and obstetrician at the University Hospital of Freiburg between 1797 and 1829. When he heard about the vaccination experiences

Johann M.A.Ecker (1766-1829) Freiburg 1797-1829

Ueber
die Kuhpocken
und
deren Einimpfung,
ein mehr als wahrscheinliches, leichtes
und gefahrloses Mittel gegen
die Kinderblattern
für
Freyburgs und Breisgaus
Eltern.

Herausgegeben und auf seine Kosten ausgetheilt
von
Dr J. A. Ecker,
ord. öffent. Professor an der vorst-rr. Albertinischen
hohen Schule, mehrerer gelehrten Gesellschaften
Mitgliede, d. Z. der wohlöbl. medicinischen
Fakultät Dekan und Protomedikats-
Verweser der Vorlande.

Freyburg im Breisgau,
gedruckt mit Felner'schen Schri.ten.

1 8 0 1.

Figure 1. Johann M.A. Ecker.

Figure 2. Lorenz Oken.

against smallpox conducted by Edward Jenner, he set out to start the first vaccination trial in Germany. In 1801 he published on his own expenses a little leaflet about smallpox vaccination in the area of Freiburg and he obtained informed consent of parents from over 1000 children in this area. He vaccinated them and a year later published the successful results again in a privately sponsored booklet on safety and efficacy of this vaccination. Ecker was also the tutor and "doctor father" of Lorenz Oken (Fig. 2) born in the nearby community of Lahr. Lorenz Oken wrote his doctoral thesis in 1803 on "Übersicht des Grundrisses des Systems der Natur-philosophie" ("Survey of the principle models in philosophy of nature"). Later on he became a leading figure in European science of the early 19th century. He became professor of biology at the University of Jena before he was nominated founding rector of the University of Zurich [1].

The next great name of the Freiburg Medical Faculty is Adolf Kussmaul (Fig. 3). Actually one might consider him the founding father of German internal medicine. He coined the unmet "categoric imperative of medicine": "Klar denken, warm füh-len, kühl handeln" ("clear thinking, warm feeling, cool acting") [2]. Kussmaul was also the founding father of clinical immunology. He published in 1866 together with the pathologist, Rudolf Meyer, the first comprehensive report on an auto-immune disease, i.e. periarteriitis nodosa (Fig. 4). This paper was only four years ago translated into English and experiences since then a great success in the United States [3].

When Kussmaul run the Department of Internal Medicine (1863-1876), a young pathologist, named Paul Langerhans (Fig. 5), doctoral fellow of Rudolf Virchow, came from Berlin to Freiburg, to write his "Habilitationsschrift". He had just finished his doctoral thesis on pancreatic islets, when he turned to a new topic: "Nerve cells of the skin" In his Habilitationsschrift published in 1871 in Freiburg he describes for the first time the dendritic cell of the skin. Initially he thought these were nerve cells but later on he corrected himself and considered dendritic cells an autochtonous cell type of the skin capable of leaving the skin under certain circumstances.

In 1875 a young student named Paul Ehrlich (Fig. 6) came to Freiburg for his 4th year of medical studies and enrolled in a practical course for advanced students ("Selbständiges Arbeiten für Geübte"), in the Department of Physiological Chemistry. Under the guidance of Prof. Otto Funke, he started to play around with chemicals for staining cells and biological material. It turned out to become his life-long fascination. In Freiburg he discovered the mast cell and wrote his first scientific publication. Later on he went to Strasbourg where he discovered the eosinophil. In Berlin he stained Koch's *Mycobacterium tuberculosis*. The move from vital stains to chemotherapy was just a small but ingenious step, which led him together with co-worker Sachahiro Hata in Frankfurt to the invention of Sal-varsan®, the first chemotherapeutical agent used in medicine. Of course all of you know his epochal achievement, namely the formulation of the "side chain theory" explaining the production of specific antibodies (Fig. 7). This won him the Nobel prize in 1908, together with Eli Metchnikoff who had discovered phagocytosis. Paul Ehrlich's thinking influenced generations of immunologists. His "side chain theory" was a milestone for the development of active and passive vaccinations, his

Adolf Kußmaul (1822-1902; in Freiburg :1863-1876)

A.Kußmaul, P.Maier: Ueber eine bisher nicht beschriebene eigenthümliche Arterienerkrankung (Periarteriitis nodosa), die mit M. Brightii und rapid fortschreitender allgemeiner Muskellähmung einhergeht.Dt.Arch.Klin.Med.1:484-518, 1866

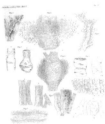

First description of an autoimmune disease: Periarteritis nodosa (cPAN) in 1866

Kußmaul's maxim:
„Klar denken,
warm fühlen
kühl handeln! "

Figure 3. Adolf Kußmaul.

Dtsch.Arch.Klin.Med. 1:484-518, 1866

XXIII.

Ueber eine bisher nicht beschriebene eigenthümliche Arterienerkrankung (Periarteritis nodosa), die mit Morbus Brightii und rapid fortschreitender allgemeiner Muskellähmung einhergeht.

Von

Prof. A. Kussmaul und B. Maier

in Freiburg i. Br.

Hierzu Taf. III—V.

Krankengeschichte.

Carl Seufarth von Gernsbach, 27 J. alt, Schneidergeselle, kam 4. Mai 1865 Morgens 10 Uhr in die medicinische Klinik zu Freiburg. An den ziemlich mageren Menschen fiel die ungemein blasse Farbe von Haut, Lippen, Mundschleimhaut und Bindehaut auf; der Puls ging sehr rasch und der Kranke fühlte sich so hinfällig, dass er sich sofort zu Bett legen musste. Er hatte übrigens den Weg ins Hospital noch zu Fuss zurückgelegt und war die zwei hohen Treppen zur innern Klinik herauf ohne Beihülfe gestiegen. Patient war einer von jenen Kranken, denen man die Prognose schon vor der Diagnose stellen kann ; er machte auf den ersten Blick den Eindruck eines verlorenen Menschen, dessen wenige Tage gezählt sind. Sein Aussehen erinnerte zumeist an chlorotische Personen, die einer rapid verlaufenden Tuberculose erliegen.

Figure 4. Report on periarteritis nodosa.

Paul Langerhans (1847-1888); in Freiburg 1871-75

Discovery of the dendritic
cells of the skin

P.Langerhans: Arch Path Anat
44:325, 1868. Habil.Schrift 1871

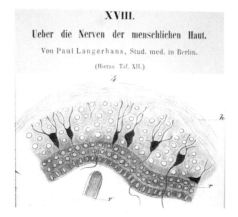

Figure 5. Paul Langerhans.

Paul Ehrlich (1854-1915); in Freiburg 1875/76

1. Discovery of the "mast cell" (1877)
2. Proposal of the "Side chain theory" (1897)
3. Introduction of chemotherapy (1909)

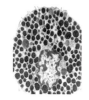

In Freiburg:

P.Ehrlich: Beiträge zur Kenntniss der Anilin-
färbungen und ihrer Verwendung in der
mikroskopischen Technik
Arch. mikr.Anat 13:263-277, 1877

Figure 6. Paul Ehrlich.

Paul Ehrlich (1854-1915)

1894-1901 — EHRLICH
Receptors for Antigen, Agglutinin, Amboceptor, Complement

Paul Ehrlich (National Library of Medicine)

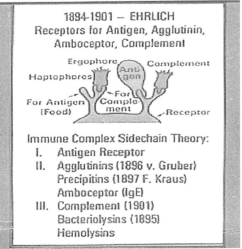

Figure 7. „Side chain theory".

concept of "horror autotoxicus" provided the most fruitful basis for subsequent research and discoveries in the field of autoimmune diseases. His foundation of the first journal of immunology ("Zeitschrift für Immunitätsforschung", today "Immunobiology") provided a forum for an expanding new science ever since.

The next personality I want to turn to is Ludwig Aschoff (Fig. 8). He was nominated head of the Department of Pathology in Freiburg in 1906, at that time he had made himself a name among immunologists with a book on "Ehrlich's Seiten-ketten-Theorie" published in 1902 [4]. Three years later he had described the rheumatoid granuloma but his greatest achievement in Freiburg was the formulation of "The concept of the reticuloendothelial system (RES)" in 1924 [5].

Between 1923 and 1936 Paul Uhlenhuth (Fig. 9) became head of the Department of Microbiology in Freiburg. Twenty years earlier (1903 and 1908) he had made an important discovery for immunology by demonstrating the organ specificity of lens proteins by means of antisera. Additionally, he showed that "sequestered antigens", like lens proteins may induce an autoimmune reaction in guinea pigs.

Unfortunately "under a ruthless Nazi government" Uhlenhuth as well as Aschoff and most of their faculty members, "lost their social tenets" (H.Krebs). In 1934 they did not oppose the relegation of their Jewish faculty members, among them so famous scientists as Siegfried Thannhauser, Hans Krebs and Rudolf Schönheimer. After the catastrophe of the 2nd World War immunology, like all other biomedical sciences, had a difficult time to get started again. Several names deserve mention for outstanding contributions to the recovery of the immunological tradition in Freiburg after the war:

Ludwig Aschoff (1866-1942); in Freiburg 1906-1935

3 major achievements in immunology:

1. Introduction of Immunology to Pathologists

1902: Ehrlichs Seitenkettentheorie und ihre Anwendung auf künstliche Immunisierungsprozesse. S.Fischer Jena

2. Description of the rheumatoid nodule

1904: Über Myokarditis.Cbl.Path 15 984/5

3. Development of the RES concept

1924: Das reticuloendotheliale System. Ergebn.Inn Med. u.Kinderhk.26:1-117

Figure 8. Ludwig Aschoff.

Ludwig Heilmeyer and his young colleague Helmut Schubothe (Fig. 10) started in 1946 a laboratory of Immunohaematology in the largely destroyed University Hospital of Freiburg. This was the nucleus of a Division of Clinical Immunology later on founded at the Department of Medicine. During three decades Schubothe was the most experienced and highly distinguished immunohematologist in Germany. He classified chronic cold agglutinine disease as a separate disease entity and wrote a textbook on immunohematology in 1958 [6].

The pioneering work on bacterial lipopolysaccharides (LPS) and their interactions with the immune system started by Otto Westphal (Fig. 11), and his co-workers Lüderitz, Färber and Fischer led to the foundation of the "Max-Planck-Institut für Immunbiologie" 1962. This milestone for German and international immunology has been the subject of historical lectures by Otto Westphal and Klaus Eichmann.

One of Otto Westphal's successors was Georges Köhler (Fig. 12). He graduated in Freiburg, wrote his doctoral thesis in Basel in the laboratory of Fritz Melchers and described in 1975 as post-doctoral fellow in the laboratory of Cesar Milstein in Cambridge a protocol how to produce monoclonal antibodies *in vitro*. This technique revolutionized immunology. In 1984 Köhler and Milstein won the Nobel prize for their epochal discovery. Tragically, Georges Köhler died in 1995 at the age of 48 years, much too early for all of us.

I hope I convinced you that the spirit of immunology has animated this place for quite some time. The legacy of the mentioned scientists has made immunology a stronghold in Freiburg. Many members of the German Society for Immunology

Paul Uhlenhuth (1870-1957); in Freiburg 1923-1936

Major achievements:

Figure 9. Paul Uhlenhuth.

1903: *Discovery of the organ specificity of lens proteins*. In: Festschrift zum 60.Geburtstag von R.Koch, S.Fischer Verlag Jena.

1910: *Concept of sequestered autoantigens:* P.Uhlenhuth, Z.Haendel: Guinea pig could be rendered sensitive to its own lens protein. Z.Immunitätsfschg 4:761

Helmut Schubothe (1914-1983); in Freiburg 1946-1983

Figure 10. Helmut Schubothe.

Major achievements:

Together with L.Heilmeyer (1899-1969):

Identification of chronic cold agglutin disease as a separate clinical entity.

H.Schubothe: Zur Frage der Spezifität, Immunologie und klinischen Manifestation von Kälteagglutininen. Zbl.Bakt.154:223, 1949
H.Schubothe: Serologie und klinische Bedeutung der Autohämantikörper.S.Karger, Basel 1958 (Monography)

Otto Westphal (1913-); in Freiburg 1951-1983)

Figure 11. Otto Westphal.

Major achievements:

1. Fundamental work on microbial lipopolysaccarides (LPS)
2. First Direktor of the MPI of Immunobiology in 1992
3. Founder of the German Society for Immunology in 1967

Georges Köhler (1946-1995); in Freiburg 1985-1995

Figure 12. Georges Köhler.

Major achievements:

Development of hybridomy technique for the production of monoclonal antibodies.
Nobelprize 1984

G.Köhler, C.Milstein: Continous culture of fused cells secreting antibody of predefined specificity.
Nature 256:495, 1975

think that the heart of our society beats in this city. We wish the "Max Planck Institut für Immunbiologie" many more years to come with lots of talented students, ingenious ideas and unlimited funds. Congratulations to the 40th anniversary !!!

Bibliography

1. E. Seidler: Die Medizinische Fakultät der Albert-Ludwigs-Universität Freiburg im Breisgau –Grundlagen und Entwicklungen. Springer Verlag Berlin, Heidelberg 1991.
2. F. Kluge: Adolf Kußmaul 1822-1902. Rombach Verlag, Freiburg 2002.
3. E.L.Matteson and H. Matteson: Polyarteritis nodosa and microscopic polyangiitis – Translations of the original articles on classic polyarteritis nodosa by Adolf Kußmaul and Rudolf Maier and microscopic polyarteitis nodosa by Friedrich Wohlwill. Mayo Press, 1998
4. L. Aschoff: Ehrlichs Seitenketten-Theorie, Fischer Verlag, Jena 1902.
5. L. Aschoff: Das Reticuloendotheliale System (RES), S.Fischer Verlag, 1924
6. H. Schubothe, L. Heilmeyer: Serologie und klinische Bedeutung der Autohämantikörper. S. Karger Verlag Basel, New York 1958.

Part III: DVD

Samples Symposium Ceremony:

Lewis Wolpert
Rudolf Jaenisch
Martin Raff
Charles Janeway, Jr.
Jacques Miller
Philippe Kourilsky
Otto Westphal
Klaus Eichmann
Hartmut H. Peter
Michael Sela